Engineering
Leadership
How to Create an Effective Engineering Organization
Robert J. Monson, PhD

Engineering Leadership

How to Create an Effective Engineering Organization

R. J. Monson, PhD

Table of Contents

Chapter 1: Introduction

This text is intended as to provide a guide for Directors and Vice Presidents of Engineering, responsible for the leadership of an engineering organization. In many cases, individuals selected for such leadership positions have been educated in a technical mindset, and have been guided largely by left-brain constructs in their career advancement. Due to this fact, many of those in these leadership positions find themselves ill equipped to manage the complexities of an engineering group, department or division. It is the development of those individuals that inspires this work.

The text is organized to cover the majority of leadership areas of interest to a technical leader in a manufacturing organization. It covers typical areas of interest to the technical leader, such as project management and Systems Engineering, but also delves into areas normally associated with industrial psychology, sociology and anthropology. These disparate areas provide insights into the appropriate management of the most important aspect of any engineering organization, the people who work within it.

These peripheral areas of development are often not included in the career plans of engineering leaders, but include important aspects of the role that cover organizational strategy, culture development, personnel management and appraisal, as well as personnel selection. These areas of development are of great importance to the engineering leader, as the proper application of these domains will greatly simplify the challenges faced within the organization.

This text also discusses the development of technology and technology road mapping, clarifying the visionary aspect necessary for success in any technical enterprise. Care has been taken to provide not only details related to the application of these concepts, but also the reasoning behind why these concepts are of importance to the leader.

The text is organized in a roughly chronological format, introducing factors of importance to the practitioner in a stepwise manner in order to achieve the maximum benefit and an appropriate introduction of new concepts. Therefore, if individuals responsible for the engineering leadership function choose to implement only the first few chapters of this text, they will receive significant benefit for their efforts. Later subjects frequently build upon the successful implementation of the earlier, requiring the foundation of strategy and a project management orientation to be successful.

Chapter 2 outlines the importance of competitive strategy and product alignment. These concepts are of significant importance to the practitioner in any engineering organization. Often it is unclear what product strategy is most beneficial for pursuit within an organization. Significant discussion surrounds the methods and approaches by which the engineering leader can define an appropriate product strategy for dissemination into the organization, and why this is of importance.

Chapter 3 discusses the value of a strong project management culture within the organization. This chapter outlines a simple and effective project management methodology that should enable any engineering organization to more effectively plan, execute and complete its projects.

Chapter 4 expands the concepts of chapter 3 from the project orientation into that of the program environment. This clarifies the differences between the project and program perspective, whereby the Program Manager must maintain a cradle-to-grave approach when dealing with the long-term developments necessary for the organization. This chapter will clarify the differences in a manner that will allow the engineering leader to better position both the project and Program Managers within the organization for competitive success with a minimum of conflict.

Chapter 5 seeks to incorporate a Systems Engineering perspective into the thinking of the engineering leader. In general, Systems Engineering is an expansion of typical project management concepts, and extends the leader's thinking in a manner that will produce a more effective and more carefully aligned product at the end of the development cycle. While many of these concepts originated in the defense industry, it is clear that they are useful to virtually every engineering leader in every organization.

Chapter 6 deals with the second type of resource planning necessary in an engineering organization. The first is created with appropriate scheduling of resources to accomplish planned tasks, and is defined in the project management methodology. The second type of resource planning includes the balancing of existing resources and existing assignments across the organization in order to identify staffing needs and challenges. A method is presented to enable the engineering leader to accomplish this resource planning approach.

Chapter 7 focuses on the development of an engineering organization that includes systematic engineering titles and career paths that enable the engineering associate to identify development needs and approaches in order to manage their own career.

Chapter 8 focuses upon the design of the engineering organization, based upon the critical performance requirements of the organization and those factors that most impact success. Proper attention to organizational design can simplify the management of the engineering organization significantly, reducing inconsistencies and rework.

Chapter 9 focuses on performance appraisal systems and the performance management of personnel within the organization. This topic is of critical importance to the engineering leader due to the necessity to provide practical and reasonable feedback to associates, which if done properly will typically increase job satisfaction, fairness and the productivity and capability of the engineering organization.

Chapter 10 suggests an appropriate business rhythm and reporting structure to be utilized in an engineering organization. This rhythm will typically produce a structured behavior within the group that will maintain project and program status in a manner that allows for full and complete disclosure within the reporting chain.

Chapter 11 delves into the Earned Value Management System, a very effective system utilized to track the current status of projects and how it may be used in a predictive manner to establish typical project outcomes prior to completion.

Chapter 12 focuses on the development of Technology Roadmaps, which when used in conjunction with product roadmaps can help provide a visionary development path for any technology organization.

Chapter 13 focuses on personnel selection within the engineering organization. By utilizing structured interviewing and a careful alignment of individual behaviors as well as organizational constraints, the engineering leader can hire highly effective employees and maintain a high-performing engineering team.

Chapter 14 deals with the ever present need to provide effective training and development for the members of the engineering staff. Several different areas of development are suggested and different methods are proposed relating to providing development in-line with engineering associates' career gaps.

Finally, Chapter 15 deals with integrating a complex system of this type in a manner that will ultimately yield a highly effective engineering organization.

Chapter 2: Competitive Strategy and Product Alignment

In most organizations, the engineering leader has limited resources available to accomplish the work required. As plans go awry and changes pop up within the organization such as resignations, sickness, and maternal or paternal leave, exquisitely detailed schedules become less relevant or even blatantly incorrect. Departmental plans become derailed, and engineers find themselves buried beneath so many demands that it is overwhelming. Many engineering leaders must accept that they have not given much thought to the complexities of multiple project environments, and instead rely upon the heroic efforts of many to keep the department afloat. In addition, many leaders do not communicate their resource planning and scheduling effectively throughout the company, and thereby often place unrealistic demands on the engineering associates. How can we possibly transform this maelstrom of changing demands, priorities and ground rules into a more stable and predictable environment? We can do so by embracing a strategy as a ground truth in the midst of Chaos, and by using the strategy to prioritize and guide our decisions.

Frequently, engineering leaders find they have significant work to be accomplished with less than adequate resources. Therefore, it is very important for leaders to most efficiently use the available resources to accomplish as much as possible. In short, given the scarce resources at hand, how does a leader ensure that they are applied to the most pressing problem or most beneficial opportunity? We utilize strategy to create focus and coherence in our decisions.

In order to completely understand the role of strategy and its value within the engineering context, it is important to consider the appropriate alignment of purpose throughout the different levels within the organization, and why that alignment may enhance or limit the organization's ability to compete effectively. We will begin by considering a simple construct of the organization, comprising three specific levels of responsibility. This can be seen in Figure 1.

The Goal-Strategy-Tactics Tasks Framework

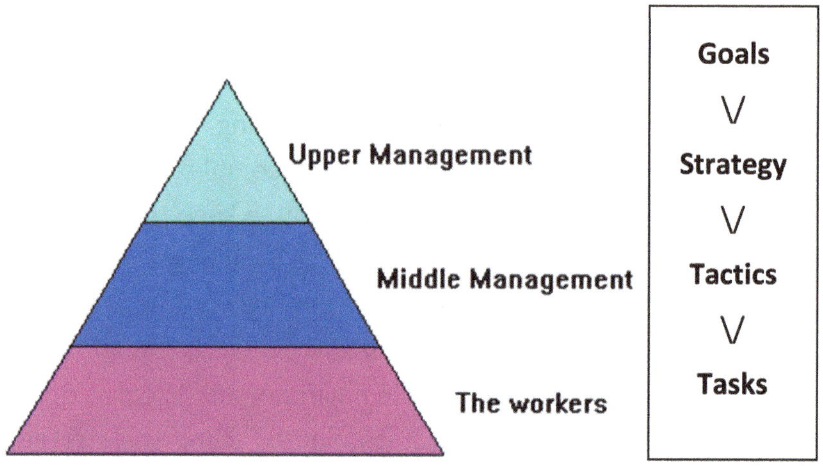

Figure 1: Goals - Strategy - Tactics - Tasks framework

At the top of the organization, we find the executives of upper management. The role of upper management within the organization is supposed to be focused upon the development of an appropriate strategy and vision for the business. This vision or strategy is intended to communicate to the other members of the organization the direction and focus, to be used instead of relying solely upon a centralized decision maker. In other words, the CEO does not need to make all the decisions; he or she wants the members of the organization to make decisions. To ensure consistency in those decisions, the CEO communicates criteria to be used for decision-making, and these criteria are equivalent to the strategy.

This concept is best illustrated when one considers the approach of centralized command demonstrated by historic military leaders when conducting their campaigns. In their world, each individual soldier comprises a single fighting unit, as well as being a member of a ten-unit squad. Normally ten squads comprised a company, ten companies comprised a battalion, and ten battalions formed a brigade. By virtue of this simple architecture, the commander is able disseminate a command throughout a large body of soldiers with relative ease. By communicating his direction or focus to the ten individuals under his direct command, who then, in turn, communicated it to the ten individuals under their command, an order could be disseminated to several thousand soldiers in a very short period of time.

One difficulty arose with this approach, however, is related to the telephone game often played by children. In this game, a phrase is whispered into the ear of the player next to you. By the time the phrase has been whispered in the ear of five or ten players in the game, it becomes largely unrecognizable due to the limitations and challenges of communications. If we extend this limitation to the previous example, it becomes apparent that the specific details of the communication might not be as readily communicated as the spirit of the communication. This is in essence what strategy is intended to convey. Therefore, upper management must focus very specifically on the development of an appropriate vision and strategy that will be disseminated (shared) amongst the organization, the intent being that the spirit of the decisions is communicated. This will effectively establish for the associates an idea of 'how' they should approach their decisions and what conclusions they should likely draw from them.

The lowest level of responsibility is that of the workers. The workers are typically located at the bottom level of the organization, and comprise the largest percentage of the population. They do not direct other workers in their efforts, and are generally responsible only for their own contributions. In many corporations, this is referred to as the individual contributor. As might be expected, the responsibilities of the individual contributor are to maximize their contribution to the organization, which is largely based on a task orientation. Naturally, this task orientation should be based upon the strategy of the organization. In other words, the individual contributors are supporting the organizational strategy by means of their decision-making throughout the business day. Therefore, the communication of the vision and strategy must be clear enough to allow appropriate interpretation by the members of the workers' group so that the leadership is assured decisions are consistent and coherent throughout all levels of the organization.

The remaining level of responsibility is that of middle management. These individuals are typically responsible for the development of projects and programs within the organization, and may also comprise the functional management of individual contributors within the organization. One is in a middle management role if he or she is making decisions that directly impact the discretionary spending of the organization in any manner that extends beyond his or her own compensation. Most Project Managers fall into this category, and virtually all engineering operates at this level within the organization. In particular, the project leads play a middle management role within the organization. As has been previously clarified, upper management's role is one of developing vision and strategy, while the workers are responsible for task-oriented efforts. Clearly the ranks of middle managers must be responsible for the connection between strategy and tasks. This implementation of the strategy is known as tactics. Thus, the middle manager must define tactics in a manner that will support the strategy while simultaneously outlining how tasks should be conducted, thereby accomplishing the necessary work of the organization.

To that end, we are interested in viewing the operation of the organization with regard to these areas of responsibility. This allows us to define a specific construct of responsibilities, known as the Goal – Strategy – Tactics – Tasks framework. This framework is defined in order to illustrate the challenges that most organizations have with the appropriate implementation of strategy. This can be seen in Figure 1.

To further convolute this issue, often the organization is split into vertical sections, or functional divisions (see Figure 2.) This will result in further separation of interests within the organization, often creating operational islands (see Figure 3.) These operational islands can be very problematic within the organization as they allow individuals within the groups an excuse to limit communication across the boundaries. While this typically is not done in a malicious manner, often the result is a costly waste of resources.

As an example, most employees in departments within an organization feel a certain affinity to their coworkers and a certain separation from those outside their department. This separation can be manifest in a variety of ways, and can include a reluctance to speak face-to-face, a limitation in communications via text or e-mail, or a perspective of an *Us versus Them* mentality within the organization. This very rarely increases the efficiency or effectiveness of products.

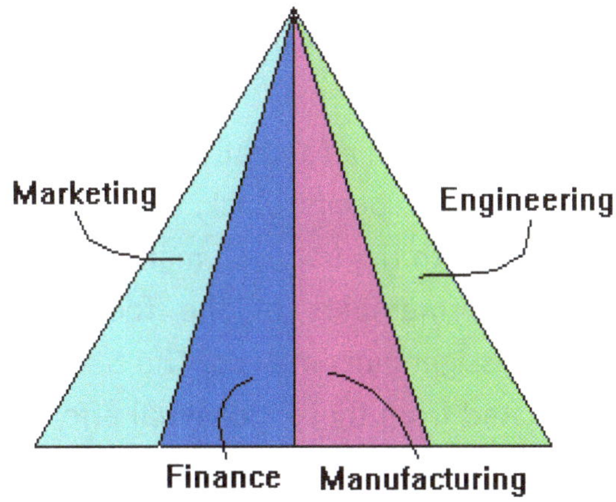

Figure 2: Functional Division

Management gaps + Functional gaps = Operational Islands => Dysfunction

Figure 3: Operational Islands

The value of the Goal – Strategy – Tactics – Tasks framework enables us to focus our efforts effectively within the organization. With proper execution, we can be sure that we have established a coherent and cohesive vision that enables the maximization of resources as they are assigned to tasks.

Imagine we want to apply this construction to a simple football game. When asked what the goal of a football game is, most individuals will respond that the idea is to win. While this is an important consideration, it is a less tangible definition that might be desired in the statement of goal. For example, when it is suggested that our real goal is to score more points than the opposition during the game, most people will agree that that is a more tangible and practical description of what needs to be accomplished by the football team.

If one considers the individual tasks within a football game, it is readily apparent that individual contributors are responsible to throw a specific block, tackle a specific player, run a pass route, throw a pass, run through the line with the ball, and many other specific tasks. These tasks are assigned individually to different members of the football team, often without direct knowledge of what all the other assignments may be. The tasks are assigned with the intention that a coordinated whole is created from these individual efforts. This coordinated whole is at the level of tactics in the football game, and it is equivalent to the offensive and defensive plays that are run during a football game.

The focus of the design of a play is to assign a wide variety of tasks to individuals on the field who are well suited to accomplish them, and then to orchestrate the whole in a manner that will allow the system to support the goal of winning the football game. With that said, it is clear that to this point we have not discussed an appropriate strategy for a football game. This is largely due to the fact that most people would approach the football game just as we have defined, without significant regard for the desired strategy of the football team.

It is of significant value, however, to consider the strategic focus for the football team. For example, if a team is highly talented at throwing a long pass, it is clear that the team would choose plays that would enable them to throw a long pass. Likewise, if a team was remarkable at running the ball, the team will choose to develop plays that would enable them to run the ball on a frequent basis. These groups of skills or talents are often collectively referred to within our organizations as 'core competencies'. Therefore, when considering the strategy, one should be considering the core competencies within the organization.

So in consideration of our football game, one should approach the goal, to win the game, from the perspective of what the players are most skilled at doing. If we as a team truly understand our strengths, it would make sense to exploit those strengths while minimizing potential weaknesses. So we must consider what strengths we bring to the table, the specific skills or competencies that we possess that will enable us to further our goal.

Once we have defined our desired focus for the game, we have essentially established our strategy, which should be used to guide our tactical decisions. For example, imagine that we are highly skilled at running the football. We would establish a strategy for our team that would enable us to utilize our core competencies and therefore more readily play our best game rather than the other team's game. This might include focusing on a short pass attack as well as a running attack, which would enable the application of core competencies and therefore allow us to take advantage of our finest points.

To sum up our approach, we have defined a goal of scoring more points than the opposition, and in order to accomplish this we have chosen to focus on our running skills (strategy) in the football match. When we need to design a play (tactic) for our team, we will take into account our strengths and competencies in order to design plays that enable us to use those strengths and competencies most readily during the execution of the game. Therefore, once the plays have been defined they will be disseminated to the individual contributors in a manner that communicates to them not only their specific assignments on the field but also the reasoning behind those assignments. It is always true that an individual contributor who understands the 'why' of an assignment will more readily adapt that assignment to the best interests of the organization. If individuals are not given task justification, they have little or no idea how to adjust their tasks within a dynamic environment.

If we were to consider a retail example, one could consider an organization such as Wal-Mart. Walking through the same thought process, we define the same elements:

Goal: to be a profitable business
Strategy: to be the low-cost provider (Why the customer values you)
Tactics: to focus on cost reduction in everything we do

The most important of these defined elements is the goal. Often goals are expressed in a manner that does not allow the individuals to easily interpret the goal for their specific assignments. In this particular case, we have chosen to define our goal as one of remaining profitable, which enables those who interpret this goal to choose a perspective based upon their contribution to the organization. If we had suggested the goal was to increase market share, this would be much more difficult to distill down to the assignments within the company.

It is clear that Wal-Mart approaches their efforts with a focus on cost. How do we know this to be the case? One need only consider why an individual would choose to go to Wal-Mart to purchase retail items. Most consumers would readily admit that they go to Wal-Mart for one reason alone, and that is to be sure that they are paying the lowest price possible for their goods. Therefore, Wal-Mart chooses to implement a strategy of cost. You will note that the strategy is defined in terms of the value brought to the customer. Another way to consider the strategy of any organization is to ask, "Why is the customer coming to me?" Defining the reasoning that makes the organization attractive to the consumer establishes the strategy that the organization should pursue and enhance.

In defining tactics for the organization or in defining those projects that will be implemented, we must consider this defined strategy of cost. In order to illustrate this point, we will imagine that we have been assigned the task of laying out the new floor plan for a Wal-Mart store in our locale. The only thing we know as a constraint in our design is that our primary strategy is to maintain the lowest possible cost to the customer. What would this floor plan look like and how would we define the characteristics of it?

To answer this question, we must consider what actions we could take that would potentially contribute to low cost. We should then select those actions that contribute most effectively to cost; those that have the highest return on investment. We would then implement those actions as our primary tactics. For example:

The floor plan will be large – in order to take advantage of economies of scale
The stock will be located on the floor – in order to minimize material handling costs
The stock will be delivered and placed on the shelves by the vendors – to reduce material handling costs
Stocking will occur during times when customers are not in the store – to minimize impact on sales opportunities
We must design multiple loading docks, and wide aisles for the transport of goods within the floor plan by forklift or pallet jack – to minimize inefficiencies when multiple vendors are restocking items

As can be seen by this simple example, we can make our decisions utilizing the criteria of strategy in order to maintain a coherent application of resources within the organization. As many of us know from observation, this is what a Wal-Mart floor plan looks like, and now we understand why.

The Strategy of the Organization

To understand strategy, we can consider the definition of the words.

Strategy –

The science and art of employing the political, economic, psychological and military forces of a nation or group of nations to afford the maximum support of adopted policies in peace or in war; the science and art of military command exercised to meet the enemy in combat under advantageous conditions; a careful plan or method, the art of devising and employing plans towards a goal.

Tactics –

Of or relating to arrangements of order; a method of applying forces in combat; a device for accomplishing an end, showing orientation or movement directed by a specific force or agent.

These definitions fit well with our previously discussed notion of strategy as it relates to maximizing the core competencies, specifically those competencies that your consumer or customer finds most attractive in your organization.

Often there is confusion relating to the definition of business strategy as well as product strategy within the organization. In the former case, we are defining the approach we choose to take toward the customer, and the reputation we seek from that customer based upon our actions. The business strategy can typically be described with a simple taxonomy as shown in Figure 4.

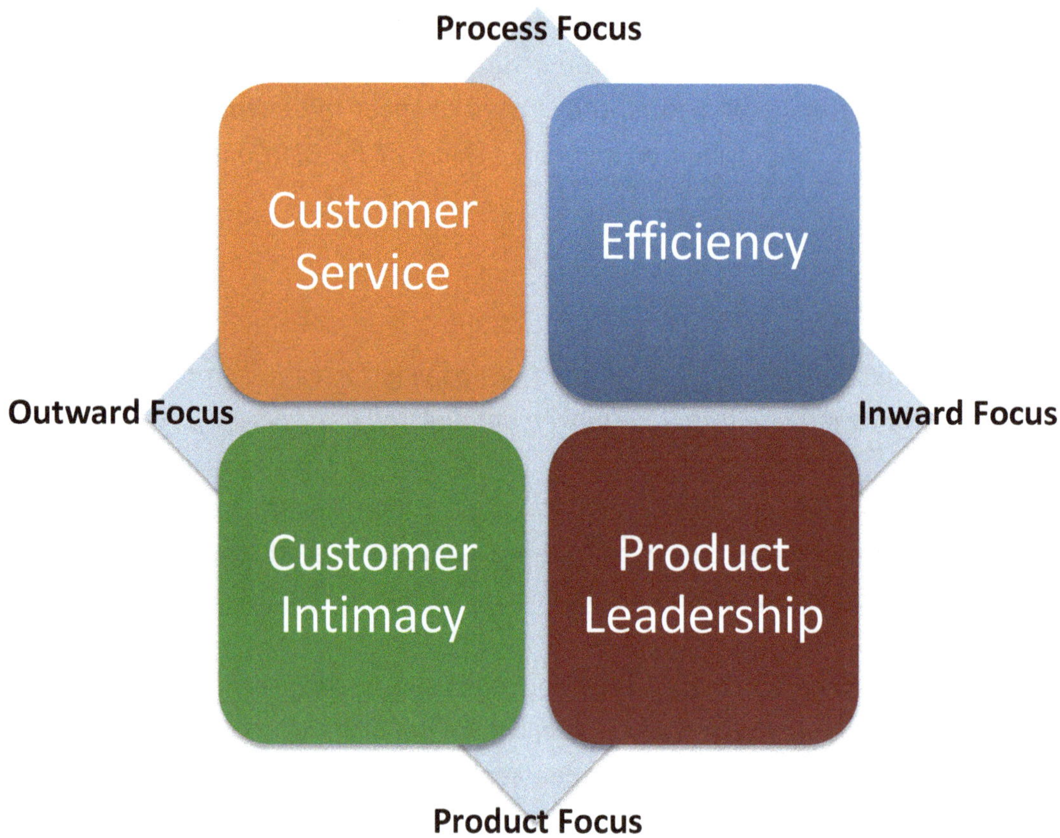

Figure 4: Business Strategy

In this example, four typical business strategies are defined and juxtaposed with regard one to another. They include Customer Service, Manufacturing Efficiency, Customer Intimacy and Product Leadership. They are arranged based upon the competitive values framework. This framework contrasts an organization's competitive interests, defining the inward/outward focus and the product/process focus present in the organizational culture. From this cultural taxonomy, common business strategies can be defined. Each of these business strategies will be discussed individually.

In their 1997 text, "The Discipline of Market Leaders," the authors describe three of these four focus areas, excluding service excellence in their approach.

Service excellence is a strategy specifically focused on satisfying customer regardless of the challenges faced by that customer from either the product or the organization. The main focus of the organization in this case is rapid response to customers, and the ready acceptance of responsibility for any shortcomings of the product is paramount. The strategy of service excellence is perfect for markets where customers value dependability of the equipment, and when that dependability is in question, support from the organization is provided to back its equipment. Leaders in this domain typically focus on having highly dynamic response teams associated with the equipment, and may limit their operations in order to maximize this capability.

Manufacturing Efficiency is a strategy that focuses on cost reduction. The typical organization applies resources to enhance its manufacturing capabilities with regard to automation and streamlining of work in order to produce the highest degree of manufacturing excellence. This strategy is typically applied by organizations that are focused on quality, as their product strategies generally require a high-volume product with highly standardized features and very little variation from those standards. This type of strategy of manufacturing excellence is ideal for markets where customers focus on cost only, which is typically the case for commodity markets and well-established technologies. This is manifest in highly standardized processes and procedures within an organization and a bureaucratic approach to making decisions.

Customer Intimacy focuses on providing whatever unique capability a customer may require for the product being offered. Often this will result in highly customized product or highly personalized product for the customer. An alternate approach is demonstrated by organizations that consistently identify the future needs of their customers, prior to the customers understanding that need. This is often defined by organizations that offer bundled solutions to problems. As might be expected, pursuing this strategy requires that the organization have a deep understanding of customer needs as well as insights into future developments that apply to these needs. This also implies that these solutions are typically not the most innovative or the most cost-effective. A dynamic development environment and an ability to rapidly adjust to changes in the marketplace will characterize an organization embracing this strategy.

Product Leadership is a common strategy employed by technology companies. This strategy is focused on possessing the superior product within the marketplace at any given time. Product leaders often can command a premium price for their product prior to commoditization of said product. Companies that utilize this approach must focus carefully on their ability to be innovative, problem solve and produce highly efficient teams. It is typically the diversity of these teams that yield the greatest results for these organizations. Therefore, companies that focus on product leadership also establish effective research management, product management and talent management programs in order to produce the visionary and dynamic workforce required to accomplish this challenging task.

While these business strategies are of significant use in focusing the efforts of organizational development, they can be challenging to apply to the individual contributor within the organization. For example, if your organization selected service excellence as a business strategy, what criteria would a design engineer utilize in order to select a bearing that provided the customer with the most effective service excellence?

Figure 5: Merging business strategies

Since it is unlikely that we will see behaviors in our organizations that require significant intellectual application or arduous tasks, such as deep thinking and interpretation of complex relationships, we must define our strategy in terms that require very little effort to interpret and apply on a daily basis. In short, our strategy must define the criteria for decision-making at the lowest levels of the organization. By limiting the breadth of interpretation available to the individual contributor, we ensure that all of the individual contributors are working coherently within the organization. This enhances our ability to compete by more efficiently utilizing resources in reducing or eliminating those squandered by misinterpretations of the strategy.

Strategy therefore provides us alignment for our team and compelling arguments for the wisdom of our direction. These compelling arguments are often very useful for the individual contributors, as they build motivation and engagement at the employee level. By having the engagement on our projects aligned with an appropriate strategy, we enhance the coherence of our efforts and assure ourselves that everyone is moving in the same direction.

Strategy essentially becomes our:

Guiding principles
Ingrained behaviors
The way we do things
Criteria for all decisions

A simple taxonomy of competitive strategies is utilized to simplify the task of strategic development. Wheelwright informs us that there are only five principal strategies typically implemented by organizations. These are known as competitive priorities or product strategies, and describe what the customer wants in your product.

Competitive Priorities

The Competitive posture in key areas of the business should focus on one primary strategy and no more than one secondary strategy. The organization needs to be reasonably competitive in all of these areas, with a particular strength or capability in the one primary focus area. This strength defines the core competencies. The five product strategies include:

1) Price	**(Cost)**
2) Dependability	**(High Value)**
3) Quality	**(High Consistency)**
4) Flexibility	**(Choice)**
5) Innovation	**(First of a kind)**

Normally a company cannot succeed at all strategies simultaneously and must choose one or two and attempt to excel in those areas, or it risks coming in at second place in the competitive arena.

A competitive strategy of price is normally associated with a focus on cost for the customer. This strategy is most commonly associated with organizations focused on high-volume production and able to establish economies of scale. All decisions focus on the bottom line for the consumer, which may also imply limited margins and profitability for organizations choosing to adopt this strategy.

A competitive strategy focused on dependability or reliability is normally associated with high-value products. Therefore the companies that utilize the strategy typically provide higher cost products within the marketplace than others' readily available products. These products typically perform at a higher level, or for a longer time, or better (in some way) for the customer. It is critical for these organizations to define that they are providing excellent value for the money for the customer. As an example, if you produce a pencil that lasts ten times as long as a normal pencil, presumably you can charge a premium price as it produces more value for the customer than a normally priced pencil.

Organizations that select quality as a competitive strategy are focused principally on high consistency in their product offerings. This is normally associated with mass production, and aligns well with a Manufacturing Excellence business strategy. Companies that utilize this strategy focus on highly repeatable products with very few changes, modifications or defects in the manufacturing process. By driving out both waste and non-value added activities, these companies can provide significant value to their customer.

Flexibility is a competitive strategy chosen by organizations that want to provide their customers with a great deal of choice. This may be illustrated by a wide variety of product models or a large option list associated with their products. This strategic choice is often of value to customers, as they are seeking highly personalized or highly focused solutions for their needs. Companies that focus on flexibility must also demonstrate an ability to rapidly adjust their product offering to the needs of a dynamic customer base. This solution often results in a higher cost to the customer and higher operating costs for the organization.

Innovation is an effective strategy selected by organizations that focus on creativity and rapid product development. Their products are often characterized by being the first of their kind in the market, new and exciting technologies, and products that cannot be sourced from any other organization. As might be expected, companies that adopt this strategy typically have high development budgets and more long-term technology development plans than organizations pursuing other strategies. They also depend on their customers' willingness to pay significantly more for an innovative product.

Implications for Engineering Leaders

It is imperative for engineering leaders to spend time considering the various product strategies that may be implemented by their organizations. Although the appropriate strategy may not always be clear within an organization, it is incumbent upon the engineering leader to define the appropriate product strategy and disseminate that understanding. A failure to do so will result in a potentially unfocused development approach, which will most likely be accompanied by failed products and projects.

One technique that has proven successful is to utilize a stakeholder analysis with regard to potential products and projects within the organization. A sample stakeholder analysis worksheet is supplied for your consideration.

Stakeholder Analysis Worksheet

Stakeholder	Overall Interest(s) in Project (Spec/Sched/ $)	What we need from Stakeholder and When	What Stakeholders want from Us and When	Key Focus for Actions

Chapter 3: Project Management Culture

Project Management skills are the least effectively taught and are the most critical for the success of Engineers in the workplace. These skills transcend the usual Engineering or Business Management skills taught to leaders, as they become life skills that can be used on a day-to-day basis and are not limited only to the Engineering domain. They have become critical because research suggests virtually all white-collar work is project-based. In addition to this work focus, most technical professionals also seek to plan their personal lives around a project orientation. Therefore, it is imperative that any effective leader in an Engineering organization demonstrates appropriate and pragmatic Project Management skills.

Many taxonomies exist that describe the project lifecycle; the Project Management Institute (PMI) adopts one of the most common. This lifecycle incorporates a five-phase approach to project understanding, shown in Figure 6.

Figure 6: The Project Lifecycle

The first phase of a project is Project Initiation. This is focused on shaping the project in a manner that will be appropriate for the individual, the project team, and the organization for which the project is being completed. This implies that the project leader is well acquainted with the organizational goals and strategy and potentially many of the tactics being implemented within the organization. With this knowledge, the practitioner can plan an appropriate project, which will fit well with the expectations and needs of the organization. This alignment is critical to ensure the effective application of resources within the organization, and often a minimum investment of time is applied to this phase. This can be problematic in later phases, as the project being accomplished may be less suitable than other projects for the future success of the organization.

The second phase of the project lifecycle is Project Planning. This is the heart and soul of most Project Managers' day-to-day existence. It involves defining the specification, schedule and budget for the project deliverables, in order to lay out appropriate resource needs, timelines and funding that are required to complete the project as specified. If a Project Manager was only to become well versed in a single phase of Project Management, Project Planning is the most suitable. With that being said, Project Planning is typically not adequate to expect project success, and the Project Manager must continue to further develop skills until a level of mastery in all phases of the project lifecycle is achieved.

The third phase of the project lifecycle and the fourth phase of the project lifecycle are typically concurrent in any given development environment. Phase three focuses on Cost Control, which is the establishment of appropriate metrics and status methods for the project. It is required to maintain updates and communications for other members of the organization, and to accurately project the likely outcome of the project at any given time. Effective Cost Control is highly useful and predictive, and it enables the Project Manager to take action quickly in the event of a downturn in project status. Additionally, utilizing a well-established methodology such as Earned Value allows easy communication across project boundaries due to a consistent and practical status system.

The fourth phase of the project lifecycle is Project Execution. This focuses on the ability of the Project Manager to engage the workforce and to enable that workforce to accomplish the objectives and deliverables set out in the project plan. This involves the application of the soft skills of management, which include teaching, learning, communication, motivation, relationship building and other related skills that are often grouped under the classification of emotional intelligence. While the project-planning phase is the most significant for the future success of the project, an ability to completely execute a project effectively is a skill not easily mastered by many technical professionals. This skill is typically not valued highly by those professionals because it requires a focus on the people as opposed to the technical aspects, something which many project professionals are ill-equipped to handle. Therefore, effective Project Management will require the effective execution of the project in addition to appropriate planning and monitoring.

The last phase of the project lifecycle is known as Project Closeout or Closure. This phase is intended to require the Project Manager to revisit the original project plans and compare them with the actual executed history of the project in order to measure the accuracy of the original projections and estimates. This assessment is intended to improve the future planning and estimating capabilities of the project leaders that will manage multiple projects throughout their career. Often this phase gets ignored when resources become tight, and project leaders move on to their next challenges. This is unfortunate, as project planners are then unable to assess their previous estimation capabilities, thereby limiting their readiness to more effectively plan and execute projects in the future. Effective Project Managers will ensure that an appropriate project postmortem and Closeout is accomplished in order to achieve these goals.

Phase 1: Project Initiation

Project Initiation is a critical aspect in the development of effective projects within the organization. Engineering leadership should take great care to assure that Project Initiation is accomplished on all projects, as this sets the stage for project alignment and the effective support of programs within the organization.

Defining Projects:

A Project is a temporary endeavor undertaken to create a new product or service (Project Management Book of Knowledge - PMBOK).

Project Management:

Project Management is the application of knowledge, skills, tools and techniques to project activities in order to meet or exceed stakeholder needs and expectations from a project (PMBOK).

Project Management Largely Consists of Two Parts:

Project Planning

The Requirements of the work

The Quantity of the work

The Resources needed for the work

Project Control and Execution

Measuring and assessing status

Making adjustments to the plan

Managing the engagement of the team

Appropriate initiation of any project requires considerable thought and reflection on the organization for which the work is to be performed. The project leader must consider what strategic imperatives are currently underway within the organization, and which competitive priorities might be most appropriately applied in the execution of the project. Once the project leader has a clear understanding of this alignment, it is appropriate to move forward in the development of the initial project plan.

To do so, one begins at the definition of the goals to be achieved by the project. These will differ from the organizational goals, but should very likely be in alignment with or in support of those business goals.

Therefore, to begin our plan we must clearly define our goals in a narrative form.

Define the overall intent of your project – The Goal

As an example, if we are responsible for the development of a new product and a chocolate chip cookie company, we must discern the ultimate goal of our project. We might be tempted to suggest that it is to dramatically increase our market share, or increase our profitability. This type of goal can be problematic, for the same reason that the application of a business strategy is challenging for an engineering organization. It is very difficult to know which recipe might be best suited to increase market share, although one might instead change that focus a bit and produce an effective goal statement for the project team.

Goal statement: Provide a new chocolate chip cookie product that will taste better than the competitors' products.

This goal statement is effective because it clearly delineates the criteria used to define the appropriate project plan to accomplish the ultimate business goal of increasing market share. Our next step, then, is to begin to define how this will be accomplished. As the reader will recall, this is synonymous with defining the strategy of the project, and how this strategy aligns with the organizational strategy. This is typically referred to as the project mission and is delineated in the mission statement.

To begin to understand how we might approach this question, we might consider additional viewpoints, such as:

How do you intend to accomplish this work? What is your general approach? What key skills and abilities will make your approach the correct one that will benefit the organization the most?

Consideration and reflection upon these questions will help to define for the project leader the competencies or skill areas of the resources that can best be applied on the project. Utilizing our example of a chocolate chip cookie company, we know that our previous strategy was to consult the finest chefs for recommendations of recipes and techniques and to use the finest ingredients available in the preparation of our cookies. Since this is the case, and we have been able to effectively apply these skills in the past, it seems appropriate for us to continue to focus on areas where we have demonstrated a significant capability or differentiator over our competition. Please also note that if the intention is to right a sinking ship, it may be appropriate to make a significant departure from past strategies in order to become more competitive or to remain competitive in the marketplace.

In this case we might choose to answer those questions by focusing on a general approach to increase the quality of ingredients that are utilized in our cookies. For example, a portion of our project may include research into what makes a cookie better tasting to the consumer, as well as research into what consumers believe defines a better quality ingredient. Once we have established this understanding we can define our mission statement for the project, which is a sentence or two describing how we will accomplish the goals. Another way to think of this is that the mission of the project should describe simply why your <u>customer</u> wants you to accomplish the project.

Project Mission: Produce an innovative cookie that the average consumer believes taste better than the competitors' cookies.

It is important to understand that this mission statement should reflect one of the five competitive priorities previously defined, including cost, quality, dependability, or flexibility for innovation. In this case we have elected to take an innovative approach to produce the best tasting cookie, which previously did not exist. This might also imply that we could charge a premium for this product.

Next, we will define a statement of work for the effort to be undertaken. This statement of work should represent an objective view of the mission to be accomplished. By objective, we mean that the final statement must be both measurable and achievable in order to establish that we have completed the deliverables we have set out to complete. Some care must be taken not to write this in a goal orientation and to assure that the final result is provable as having been completed.

Goal: A vision or intent that may or may not be achievable

Objective: A measurable and achievable goal

Statement of work: Develop and document a new chocolate chip cookie product utilizing higher milk-fat chocolate and all natural dairy ingredients to produce the best tasting cookie for our customer.

Last, the purpose of the project should be defined. This should describe why the organization is undertaking the completion of this project.

Purpose: To provide a new chocolate chip cookie product that will result in the development of a new market of high quality cookies with superior flavor for our company.

Phase 2: Project Planning

Project Planning is a crucial component of the project lifecycle. There are many different published methods regarding planning, and several will be discussed here. Fundamentally, it comes down to a very simple understanding of what needs to be accomplished in the Project Planning process. There are three fundamental questions that must be answered:

What are you going to do?
How long is it going to take?
How much will it cost?

If these three questions are answered effectively, any individual can make an appropriate business decision based upon knowledge of the organization and this scoping of the project. Therefore, we will spend our time focusing on how to address these specific needs.

In order to answer the question, "What are you going to do?" we must scope the project appropriately, provide requirements that must be fulfilled for each of the appropriate stakeholders and ultimately derive a specification describing the fit and function of the system being developed.

The specification will describe every aspect of the system under development, but we also want to identify the higher-level deliverables that must be provided by the system in order to provide a complete description of what will be accomplished on the project.

There are several approaches that can be used in order to develop descriptions of what will be accomplished, but prior to that a significant effort that must be undertaken in order to define requirements necessary to achieve all of the stakeholder's needs under the project development effort.

Requirements Definition

A requirement is something that the product must do or a quality that the product must have. A requirement exists either because the type of product demands certain functions or qualities, or the client wants that requirement to be part of the delivered product.

Why focus on requirements?

To ensure project success, 8% - 14% of project costs should be invested in requirements development
Requirements define the project scope
Make / Buy decisions can be made based upon the requirements
Requirements can define how we implement the project
Requirements reveal where further research is necessary
Requirements help define what we do not know
This allows the requirements to drive the solution - not the reverse

Reasons for Good Requirements

Save time and money in project development
Increase efficiency of communications
Set appropriate expectations and parameters for the project
Prevent duplication of effort by project team members
Focus project resources on key milestones and maintain coherence
Spend less time on revising product or service requirements during or after project implementation
Fewer contract changes with external vendors and associated costs
Better product or service quality
Performance accountability – to be able to do what we said we would do

Many different methods have been defined to enable us to begin to develop requirements. In general, there is no ideal method that will ensure that all the requirements or even the most critical requirements have been identified. Utilizing multiple techniques such as those described in the following paragraphs will be very useful in establishing a large percentage of those requirements that will be of interest to the project team and customer.

Requirement Definition Techniques

Interviews/Surveys/Focus Groups

This method is typically used when little is known about the expectations and needs of the customers and stakeholders. The basic premise of this methodology is to talk with people utilizing a structured interview technique in order to define:

What do they need from the developed project?
What do they want from the developed project?
What are they concerned about with regard to the effort?
What other elements of interest come to mind when they are considering the project?
What capabilities are implied that must be present in a successful system?

This methodology can be quite time-consuming, but allows its user to discover interesting perspectives and innovations that might not readily come to mind without the input of those most familiar with the application, the stakeholders.

Mind Mapping

Mind Mapping is a methodology that begins with our project focus located in the center of the mind map. From that center, we connect all related topics that might be of interest to us with regard to the project. This allows us to create a series of spokes around our central concept that will bring to mind many considerations we might not have perceived prior to this exercise. In addition to this, we can continue to connect spokes outward from the related ideas to more fully develop an understanding of what challenges need to be addressed in the project.

This can be seen in Figure 7.

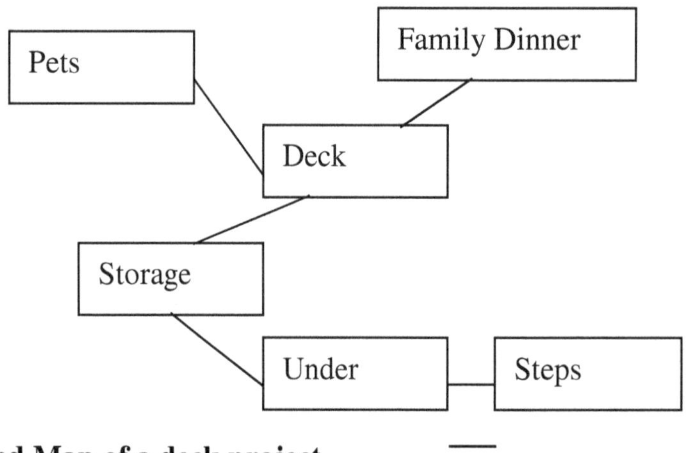

Figure 7: Mind Map of a deck project

Workflow Analysis

In workflow analysis, establishing the workflow throughout the lifecycle of the system being developed generates its requirements. This will essentially result in a Start – to – Finish Block diagram that is used to operate the system. From this diagram, a developer can carefully examine the relationships and establish whether any of the steps may be eliminated or if any challenges are present in the current workflow.

Additionally, this workflow can be connected to other systems within the organization, allowing the developer to explore the relationships amongst these interrelated systems. From the workflow analysis, each box and each connector will typically result in one or more requirements being identified.

A Cake Example produces a workflow:

Decide on Cake
Find Recipes
Check Inventories
Go to Store
Buy Groceries
Prepare Kitchen Components
Mix Cake
Grease Pan
Bake Cake
Cool Cake
Make Frosting

Frost Cake
Serve

From this example, it becomes clear that the project plan must include recipes and buying groceries in order to complete the overall project. This type of workflow analysis can help to create boundaries around the project enabling the stakeholders of the project to have a common understanding of what will be accomplished and what will not be accomplished within the project development.

Systems Analysis

Systems analysis is commonly used in complex systems development. The preferred method typically utilizes an object-oriented model of the system functionality in order to fully understand the interrelationships of the various elements of the system as well as the capabilities that must be delivered in the final product. This is similar to dataflow modeling in software development or event response modeling as is commonly used in electrical engineering.

As further detail is added to the system model, each of the boxes can be expanded into multiple functions within the box, which also generate additional understanding and requirements. In the Systems Engineering field, this type of systems analysis is what generates the lion's share of requirements for the system developer. Further details on this technique can be found in the Systems Engineering section of this text. An example of a top-level systems analysis is shown in Figure 8.

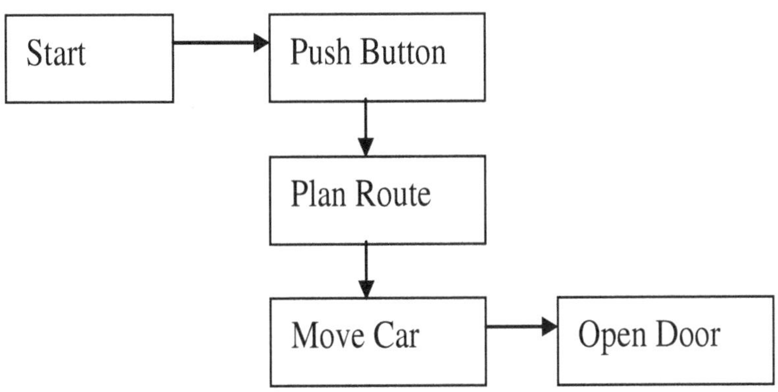

Figure 8: Systems Analysis of an elevator

Prototyping

Prototyping is undertaken in order to address one of the fundamental shortcomings of customers in the project development process. The problem is in specifically identifying what they need or want from the project. Often this is due to an inability to see what is currently not existent, and it limits the customers' perceptions only to what they already have. The use of prototyping can generate new products and new approaches that will enable customers to better understand what you intend to accomplish and may provide additional requirements based upon their observations regarding the prototype.

The prototype can be a display-oriented or functionally oriented device, and can be rudimentary or elaborate depending on the technology availability and the criticality of the prototype results. In some cases extremely high fidelity prototypes can be created in CAD systems in order to provide the customer a realistic view of what the finished product will look like.

Additionally, prototypes may be used where modeling or simulation could become too demanding or expensive in the development of a usable end product.

Scenarios / Use Cases

Scenarios are activity-based descriptions or models of the system's intended use. Use cases are similar to scenarios except that they also include the actors associated with the activities. Both of these approaches might be very useful in understanding the interrelationships between the systems as well as between the actors in the system in a typical scenario. Additional insights and requirements may be generated for the system by walking through all of the activities and relationships.

Market Research

It is often of benefit to the project leader to conduct significant research into the market or the project's technology area of interest. This research might take the form of:

Internet searches

Publications review

Reviews and discussions with business contacts

Many project leaders are surprised to discover that in many cases other parties, sometimes in unrelated industries, have already accomplished the work they are contemplating. Some research is typically required to better understand the challenges faced by the project team prior to project launch.

Templates/Project Warehouses

In many cases it is well worthwhile to examine past efforts undertaken by the organization or records available in the public domain from other organizations that have accomplished similar projects. These documents can describe standard requirements that are utilized in that industry, and save considerable time in requirements development. These documents might include:

RFP – request for proposal*
Project introduction/kickoff documents
Business objectives
Background information supplied with the project
Scope descriptions from project sponsors
Business terms and conditions from contracting agents or vendors
Requirements documents included in business agreements
Proposal instructions included in an RFP
Evaluation criteria included in an RFP

*Note that this document is usually not available at the project level; this is more of a program level activity.

Performance Based Requirements

Performance-based requirements are typically the most useful for the project development team. These are requirements that are measurable and achievable where the performance measure is included in the description of the requirement. This requires that care be taken in the development of requirements to ensure the resulting requirement description is appropriately constructed and has enough detail to provide a performance-based profile.

Several rules are utilized to generate effective requirements. These include:
1) The requirement statement is a complete sentence.
2) The requirement statement uses the word 'shall' in its construction.
3) The requirement is limited to a single thought.

Example performance-based requirements will focus on issues such as cost, quality, speed, efficiency, weight, and size to name a few.

Example performance-based requirements might include:

General: Must print fast.

Performance-Based: Must print three inches per second.

General: The software system shall retain history.

Performance-Based: The system shall retain a minimum of sixty weeks of history and a minimum of four weeks of schedules.

General: Training shall be provided.

Performance-Based: Three training sessions shall be provided to thirty employees onsite at XYZ Corporation.

Types of Requirements

Functional - What does the product or service need to do?
Action requirements
Workflow steps
Examples:
Sum total hours worked
Calculate payroll
Print paychecks

Technical - Conformance to standards
Conformance to IEEE 802.11(b) Standard for radio frequency
System type
900 MHz vs. 2.4 GHz direct sequence spread spectrum (DSSS) system.

Usability - Describes how the functional requirement should be perceived
User-Friendly
Graphic user interface
Wizard-driven screens
Reliability: Mean-time between failures
Speed: Number of seconds/minutes/hours to complete
Accuracy: Percent accuracy or allowable errors
Number of steps to complete

Operational/Environmental - Describes the system environment
Components must tolerate external temperatures from –20 degrees to 110 degrees F.
Equipment must withstand 95% humidity without deterioration in performance as measured by MTBF

Security
Confidentiality
Encryption, authorization, authentication, privacy policies
Accessibility
Legal considerations

Physical characteristics
Equipment size limitations
Portability
Weight limitations
Materials – durability
Company brand requirements

Support
Maintainability
Instruction manuals if to be maintained internally
Knowledge articles for internal support centers
Spare parts
Developer kits
Service Level Requirements
Repair turn-around (e.g. Must be repaired within twenty-four hours of service call)
Service representative availability (e.g. 24x7)

Training
Train the trainer
Train specified number of business or technical users
Describe level of participant skills
Number of hours/days
Specify content to be covered at a high level

It is generally considered good practice to ensure that some requirements appear in each of these specific areas after the requirements definition is complete. Once all appropriate requirements have been defined, a detailed statement of work and specification can be completed. An example statement of work document has been included for reference.

Statement of Work Worksheet

Project Title:	
Project Leader:	Date Written:
Project Purpose:	
Background:	
Project Description:	

Deliverables:
Measures of Success:

Alternate Statement of Work worksheet

Project Name:

Project Mission (Strategy):

Objective Statement:

Specification:

Project Specifications

The development of a detailed specification is intended to convey to all interested parties exactly what will be accomplished on the project, and what the characteristics of those deliverables will be. This is essentially how we will satisfy all the needs and requirements through the development project. This will also support the development of the top-down deliverables for the project.

The specification is intended to convey, 'What we will do', and not describe how it will be done. In general there is a very fine line between these two points, but it is important to draw a distinction between functionality and implementation.

Specifications should be as concise as possible, but should also effectively communicate the look and feel of the project deliverables, the functions or services provided by the deliverables, and the key performance parameters (KPP) for the project. The specifications should also convey the success criteria for the project.

To develop a specification one can utilize several techniques.

1) Consider the high-level project deliverables:
    - Define each tangible and measurable component
    - Define the components' notable characteristics
    - Establish several requirements in each of the domain areas

2) Establish the functional hierarchy
    - What is the top-level deliverable?
    - From what subcomponents may this be made?
    - What elements comprise the subcomponents?

Imagine we are defining a specification for a vacation we are planning in the coming year.

What are the deliverables?

An enjoyable vacation

Subcomponents of this include:

Travel to the site
Housing at the site
Meals

Entertainment/chores
Travel back home

Our specification should reflect each of these areas.

Our next task is to add additional details to each of these subcomponents as follows.

Travel to the site

Plan from home to the airport
Plan for a flight to the destination locale
Plan for transport from the airport to the destination
Established luggage plan for all of these
Consider meals as required during the transport

Housing

Plan for seven nights
Plan to remain at the same hotel
Make sure the quality of lodging will meet our standards

Once this is completed, we can continue to add additional detail in order to complete our specification.

Project will include travel to the destination site
Including:

Reserved cab ride to airport for a party of four
Reserved flight from MSP to ABC for a party of four
Reserve midsize rental car at ABC for eight days
Luggage to be limited to one carry-on per member of party
Luggage will be checked from MSP to ABC
Meals will be purchased at MSP and ABC for party of four

Project will include housing at the destination including:

Reserved two rooms of double occupancy at four-star hotel
Seven-night stay
Etcetera

Simply by increasing the level of detail included in the description, a thorough specification can be developed. In general, detail should be added until the project becomes specific enough that it couldn't be misconstrued by the majority of readers of the project plan. While it is true that some of the individuals associated with the project may purposely misconstrue elements of the specification, it is unrealistic to expect an all-inclusive specification that will provide an airtight description of the project to be completed. The final project specification should take the form of a written document describing the requirements and specifications for the system to be developed. This may look like the following:

Requirements for a household toaster (partial list)
The toaster shall operate on 120 VAC 60 Hz power
The toaster shall allow for user adjustment
The toaster shall have a power cord utilizing an LG plug
The toaster shall be well tested
The toaster shall provide indicators for the user
The toaster shall have maintenance access panels
The toaster shall be portable
The toaster shall contain a safety override ground fault circuit interrupter

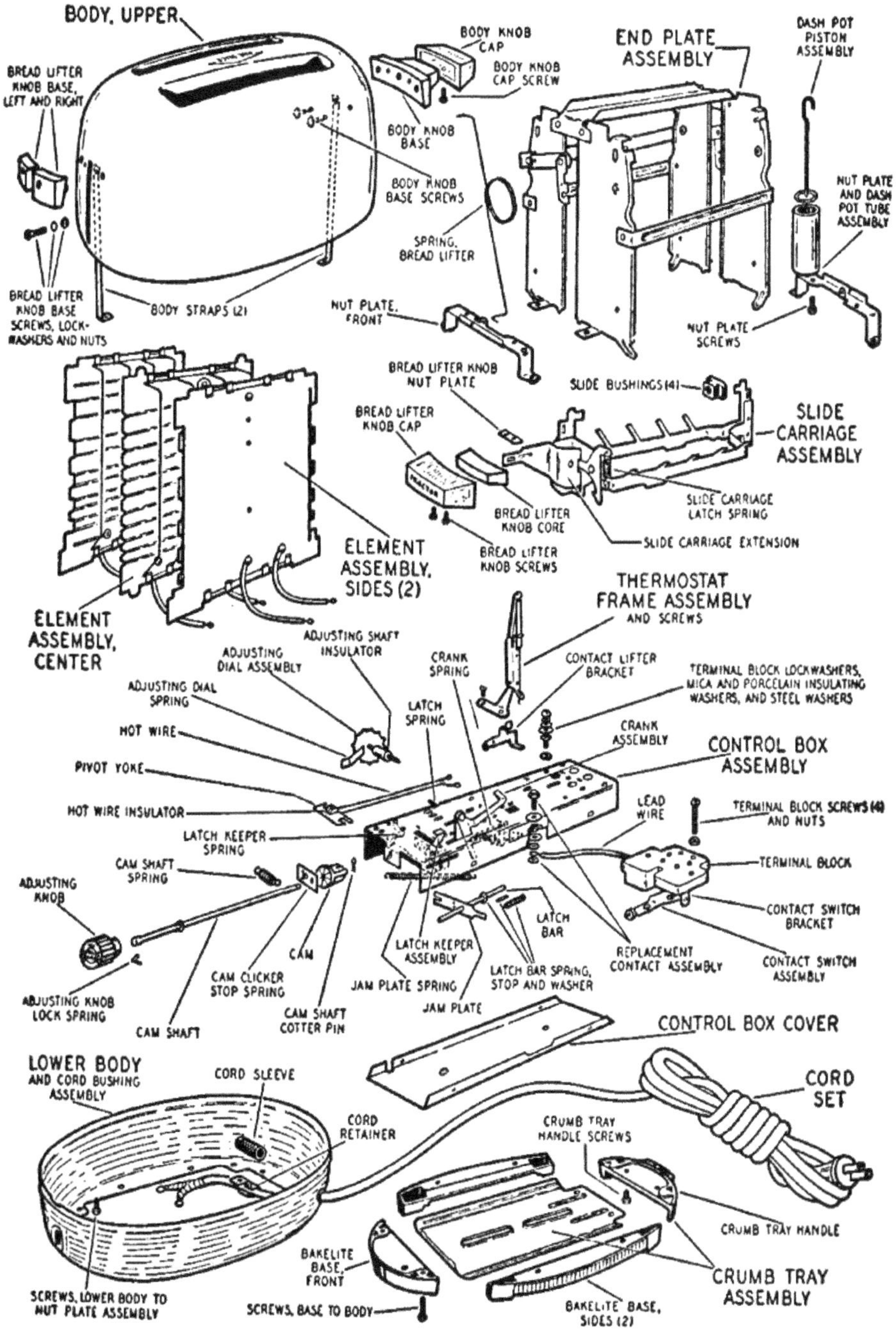

Figure 9: Graphical Toaster Specification

At this point we have defined one of the three factors of interest to the Project Manager; namely, 'What we will do'. Next we move on to the development of the project schedule.

The Project Schedule

In order to develop a project schedule, project leaders must shift their thinking from the tangible deliverables of the project to a hierarchy of tasks that must be completed in order to accomplish the needs of the project. Different methods are suggested for the development of this hierarchy of tasks. The author prefers the approach of Functional Analysis to define not only the capabilities of the system but also for the identification of the tasks that must be accomplished in order to produce those capabilities.

To conduct a Functional Analysis, the project planner must consider the operation of the system under development, and identify all needed capabilities by means of a verb/noun pairing. This is intended to provide a series of capabilities that must be assembled into a holistic view of how the system shall operate. This can be accomplished by a free association exercise in which all of the capabilities are identified, allowing time later in the development to fill in any missing functions or capabilities.

If we consider a simple example, the development of a countertop toaster for our home, one can imagine that we have identified a variety of capabilities that are communicated in this format:

Create setpoint
Clean cycle
Convert power
Update software
Override heat control
Shutdown power
Apply energy
Sense width
Toast product
Since complete
Shut down heat
Unload product
Loan product
Signal status
Input command
Request update
Reset manually

Clean device
Check load condition

These functions must then be assembled into a practical whole, which will provide an Operational View of the toaster system. This is shown in Figure 10.

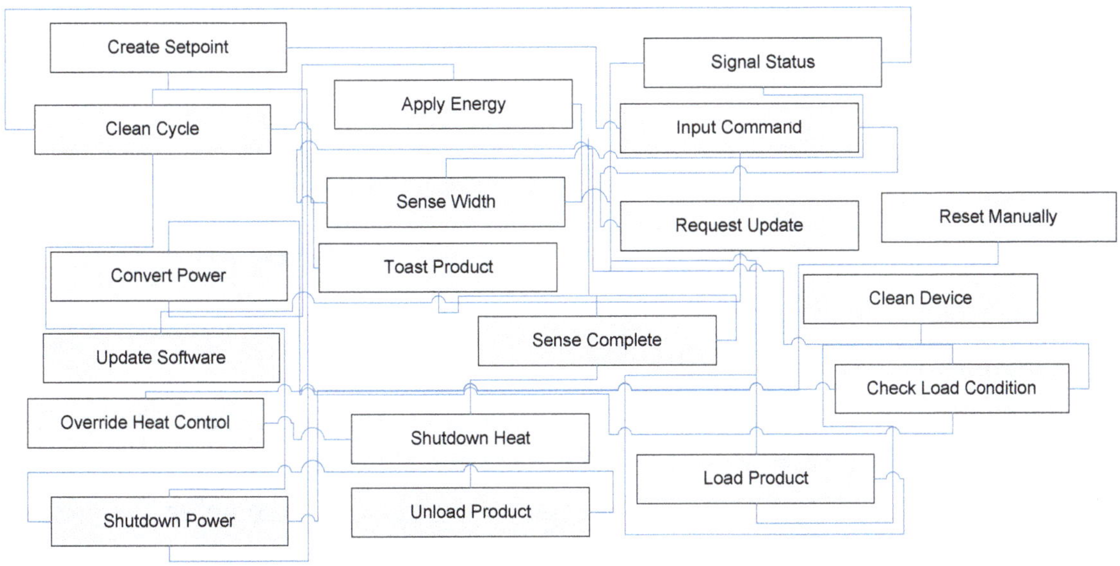

Figure 10: Functional Analysis of Toaster

Once all of the functions have been related one to another, the planner can untangle this diagram, producing what is generally constructed as a top left to bottom right flow of operations for the system. At this point it might become clear that there are several shortcomings that must be addressed, and additional functions may be added or superfluous functions may be deleted. The final outcome is to produce an understandable flow of operations for the proposed system. This is shown in Figure 11.

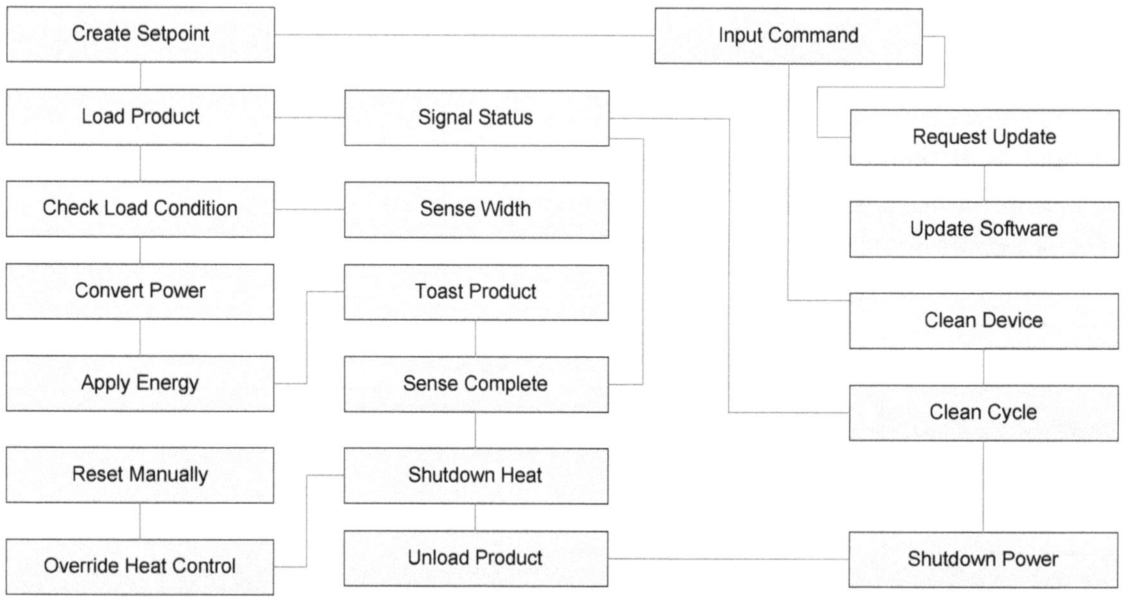

Figure 11: Simplified Functional Analysis

Once the simplified Functional Analysis has been diagrammed to the satisfaction of the project planner, these functions may be grouped into similar functional families, which will ultimately be used to define the major subcomponents necessary to successfully accomplish the project. This is shown in Figure 12.

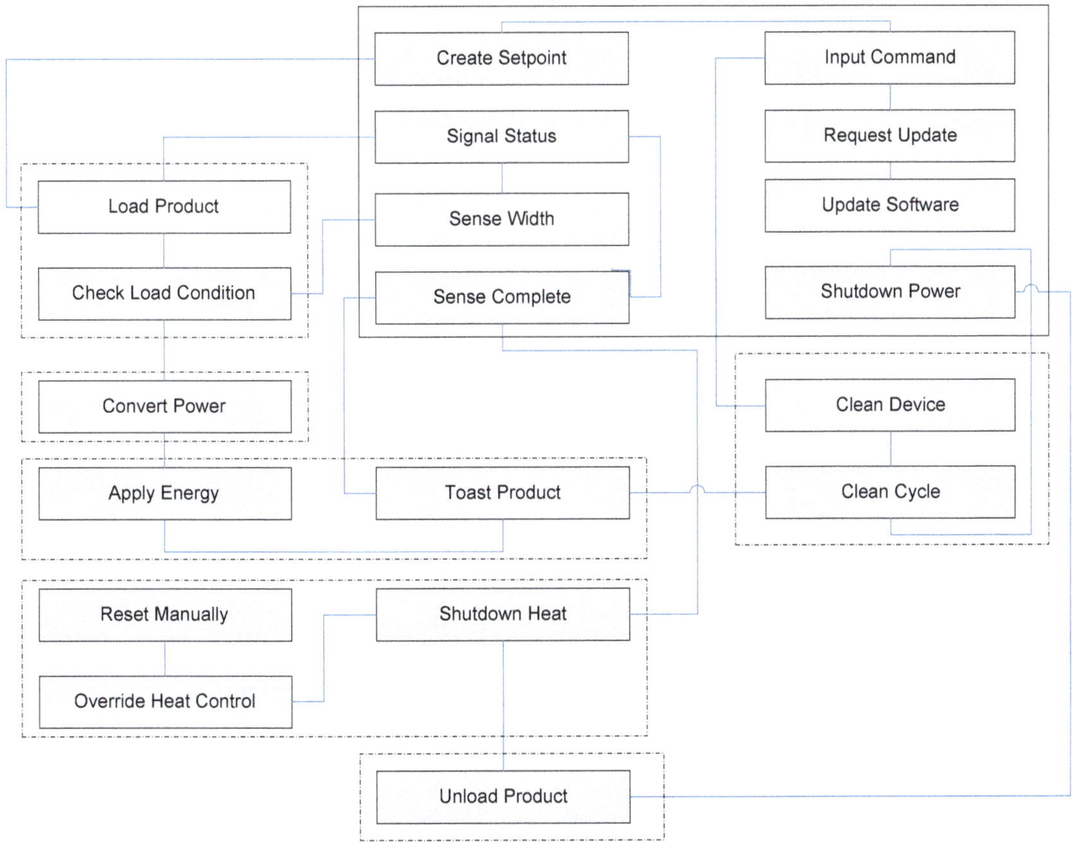

Figure 12: Grouped Functional Analysis

Once the Functional Analysis has been grouped, these groups can be collapsed into collectives or subcomponents that must be provided by the project. This is of particular importance as it defines the major subprojects that will be implemented during the execution of the project, and will coincide with the major headings in our hierarchical task listing. As can be seen, we have identified the top level of our system, the first level of indentation and the second level of functions that must be generated in order to complete the project. This construct will be extremely useful in the development of our work breakdown structure (WBS), which is the list of hierarchical tasks needed for the completion of the project. The WBS is also the most significant input to our schedule. This is shown in Figure 13.

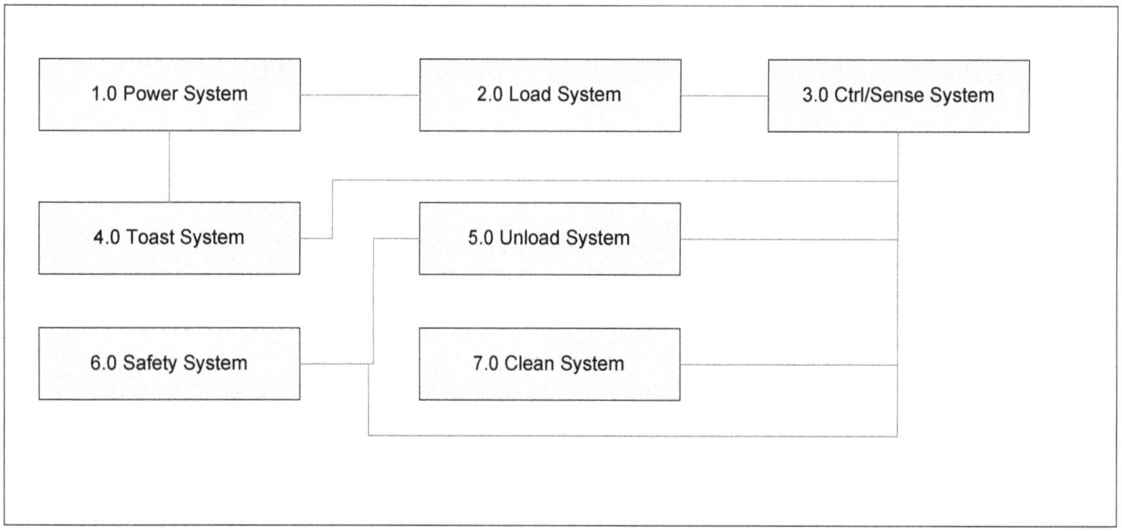

Figure 13: Subcomponents of Toaster

Once this effort has been completed, we are ready to begin the development of our hierarchical task list. This list will be utilized as an input to our schedule to define all necessary tasks that must be accomplished on the project. To clarify for our toaster project:

Toaster project

- Power system
- Load system
- Controls and sensing system
- Toasting system
- Unload system
- Safety system
- Cleaning system

To begin the generation of our work breakdown structure, we must translate these needs into tasks:

Toaster project

- Develop power system
- Develop load system
- Develop controls and sensing system
- Develop toasting system
- Develop unload system
- Develop safety system

Develop cleaning system
Integrate systems
Prototype toaster
Test toaster
Release of System

Note that additional tasks have been added to the list to accomplish the integration and testing of the system being developed. The next step is to add additional details to each of the subprojects in order to describe all those tasks necessary to accomplish that milestone. For example:

Develop power system
- Layout power system
- Generate electrical drawings
- Generate wiring diagram
- Generate installation drawings
- Generate bill of materials
- Create documentation package
- Release power system

Additional detail can be added to each of these indented task groupings to create a completed hierarchical task list. Additional detail is typically added to the task list until the final task descriptions become self-explanatory to the reader. It may require many levels of indentation, but the further we delve into the project schedule, the more robust our project plan becomes.

We can see this in an example of making a chocolate cake:

Chocolate cake project
- Planning
  - Locate recipes
    - Select cake recipe
      - Select Grandma's chocolate cake recipe
    - Select frosting recipe
      - Select Four Seasons' chocolate glaze
- Logistics
  - Reserve kitchen for Saturday

Clear calendar for Saturday

As can be seen, the same technique is utilized in order to establish the entire hierarchical list of tasks. This is also known as the work breakdown structure, or WBS. Once this list of tasks is complete, we can incorporate it into a variety of project planning software packages. This WBS is inserted into the task description column, as shown in Figure 14. This format of project plan is known as the Gantt chart, and is characterized by a hierarchical list of tasks associated with a calendar.

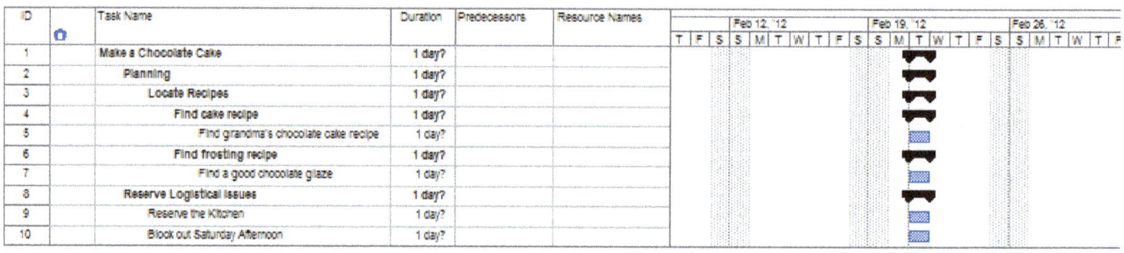

Figure 14: WBS input to Gantt chart

Note that on the Gantt chart, there are ten independent tasks. Of these ten tasks, only four are shown as bars in the diagram. These are also coincidently the tasks that are indented the furthest to the right throughout the project plan. As it turns out, these are the only tasks that require work to be completed. All other levels of tasks that have been out-dented from these tasks are simply summary tasks and are used as placeholders in the project plan (they define the hierarchy).

What this means is that we have only four tasks or work packages that must be completed to complete this scheduled project. Software packages such as Microsoft project and ProjectLibre take care of this identification automatically for us.

Our next challenge is to identify the relationship between the various tasks on the project plan. This is normally accomplished by linking the tasks one to another in order to clarify the order in which these tasks should be accomplished. The most common method links the finish of a task to the start of the following task. Other methods such as the start of a task may be linked to the start of another task as well. Ultimately, this linking enables us to establish the total duration of the project. This is shown in Figure 15.

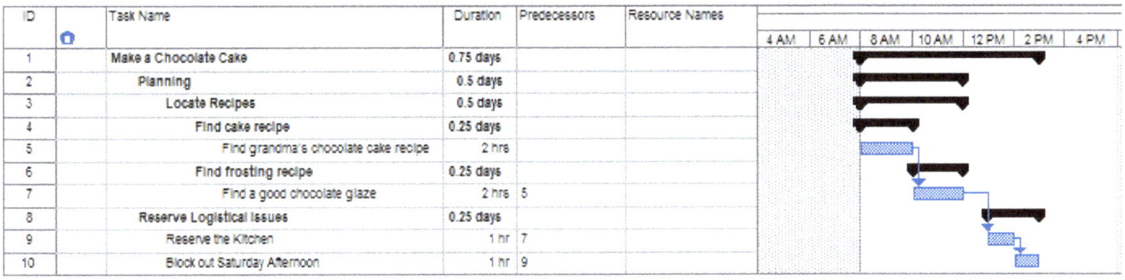

Figure 15: Linked project schedule

Durations for each of the work package tasks must be defined at this point. These are input into the Gantt chart, and result in a chronological view of the project to be completed. Only the work package tasks are assigned durations and linked. All other tasks are summary tasks and simply follow the positioning of the work packages.

Durations are estimated based upon the planner's knowledge of the work to be completed. In cases where the planner does not have sufficient knowledge to feel comfortable with these estimates, the aid of a subject matter expert is used. In some cases, if completed project plans are available with actuals that are similar to the tasks being planned, those actuals might be used as a basis of estimate.

At this point, resources are assigned to the work packages in order to establish what skill areas (and therefore what billing rates) are necessary to accomplish those tasks. This is shown in Figure 16.

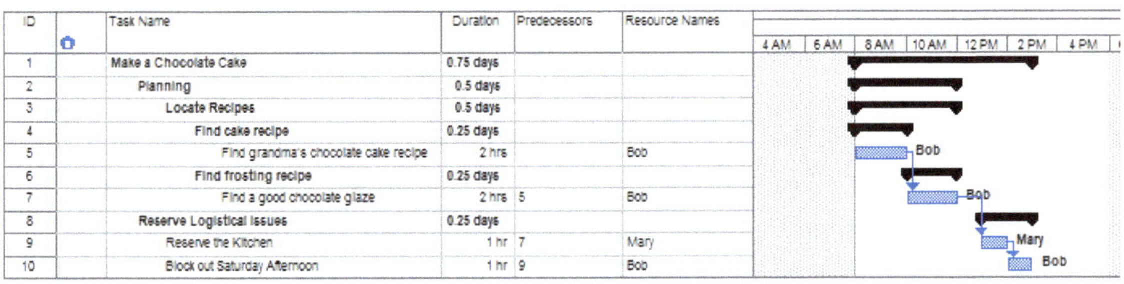

Figure 16: Resource Assignments

Resources are only assigned to the work packages. This ensures that resources are only paid for once in the project's final roll up of the budget. If resources are added to the summary tasks, the project leader will pay for those resources multiple times during the execution of the project.

Indentation levels of the task descriptions are related to the hierarchical task construction, whereas the indentations on the schedule portion of the Gantt chart are chronological in nature. These levels of indentation are entirely independent of each other.

Utilizing this methodology, we have successfully generated a detailed project plan to make a chocolate cake. There are multiple methods that can be used to accomplish this, and any means by which the project planner can generate an effective work breakdown structure and Gantt chart are considered reasonable in the generation of the project plan.

Networking method

A second method of plan generation is the networking method. Imagine you have been provided a list of tasks to be completed, in no particular order. Begin by assigning a letter to each of the tasks:

(A) Bake ham
(B) Make sauce
(C) Warm buns
(D) Cut ham
(E) Preheat oven
(F) Buy food
(G) Get recipes
(H) Make salad

Build a network of the tasks based upon their interrelationships with each other. Then assign a duration value between the tasks representative of the effort required for completion. This is an event-based system, and allows easy modification of the network as tasks are added or deleted. This network is shown in Figure 17.

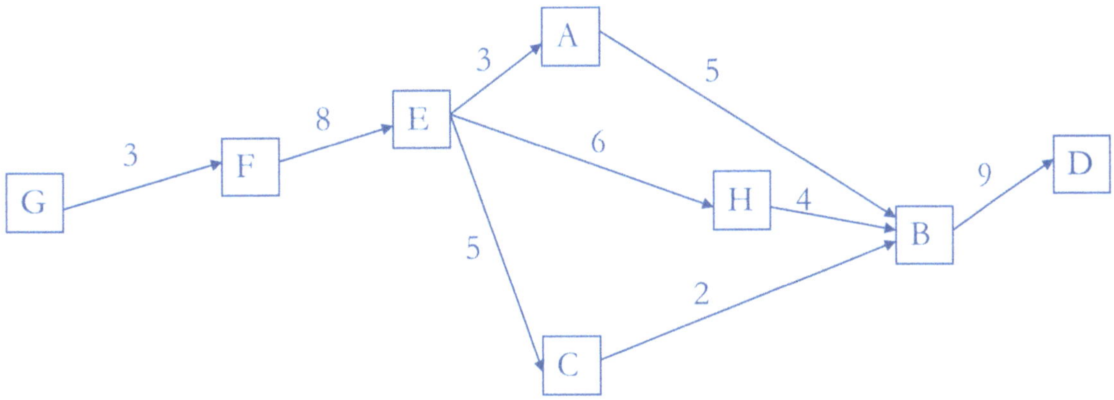

Figure 17: Network diagram

The network diagram has a significant advantage in that it can help us to identify the critical path. This critical path is of significant importance to us as it represents the longest path from the start of the project to the end of the project. Those tasks that fall along the critical path are critical due to the fact that any delay in the critical path results in a commensurate delay in the delivery of the project. The critical path in this case is shown in Figure 18, and is comprised of tasks G-F-E-H-B-D.

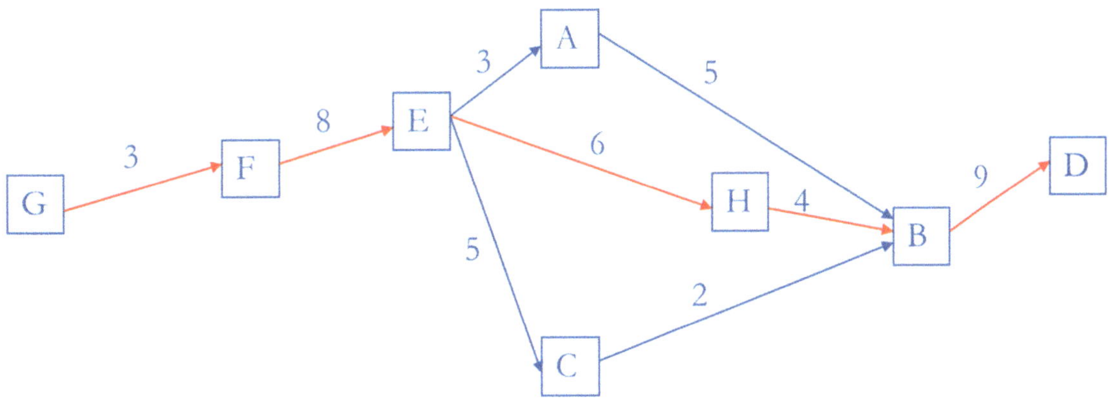

Figure 18: Network with critical path

In this instance the critical path contains ten units of time between tasks E and task B. The alternative paths require eight and seven units respectively. This means there are several units of float associated with each of these pads, which actually means that the eight units of work may float anywhere within the ten units allotted and still finish without impacting the completion of the project.

The network method of scheduling is often used by project planners due to the perception that it requires less work than the production of the Gantt chart. In reality these two approaches contain the same information for the project plan. In fact, if you choose to transition from a

Gantt chart to network chart, the methodology is quite straightforward. First we will begin with the Gantt chart, as shown in Figure 19.

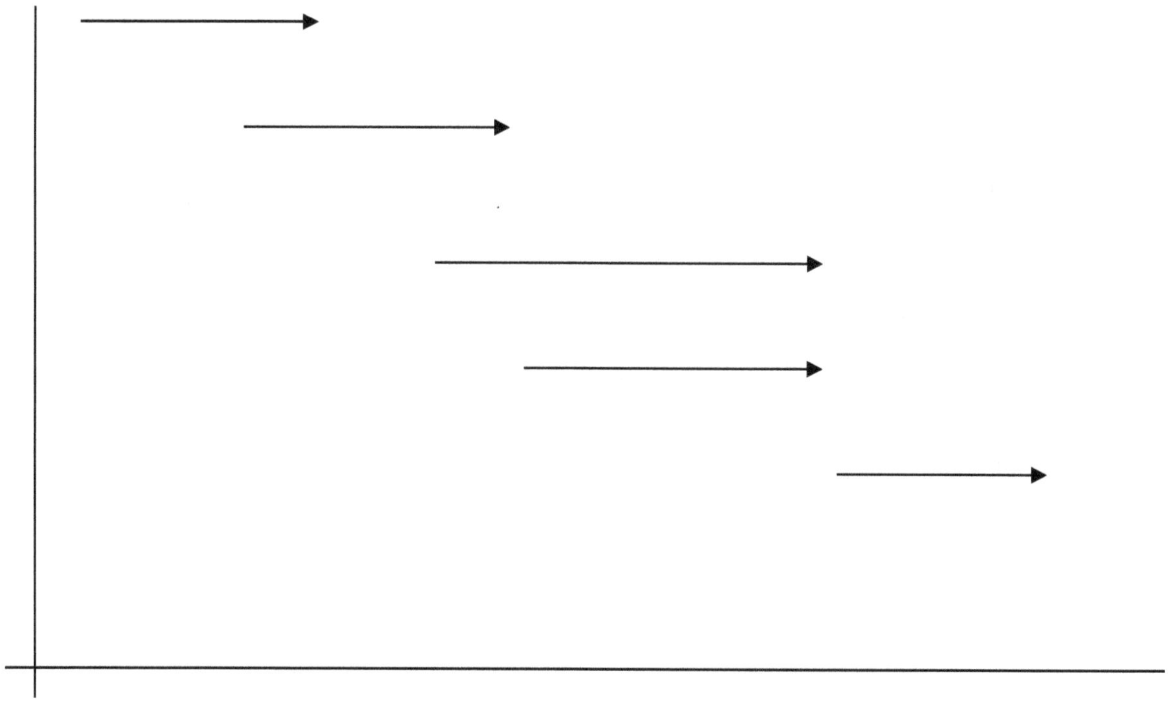

Figure 19: Gantt chart

Next we will assign an event to the start and finish of every task.

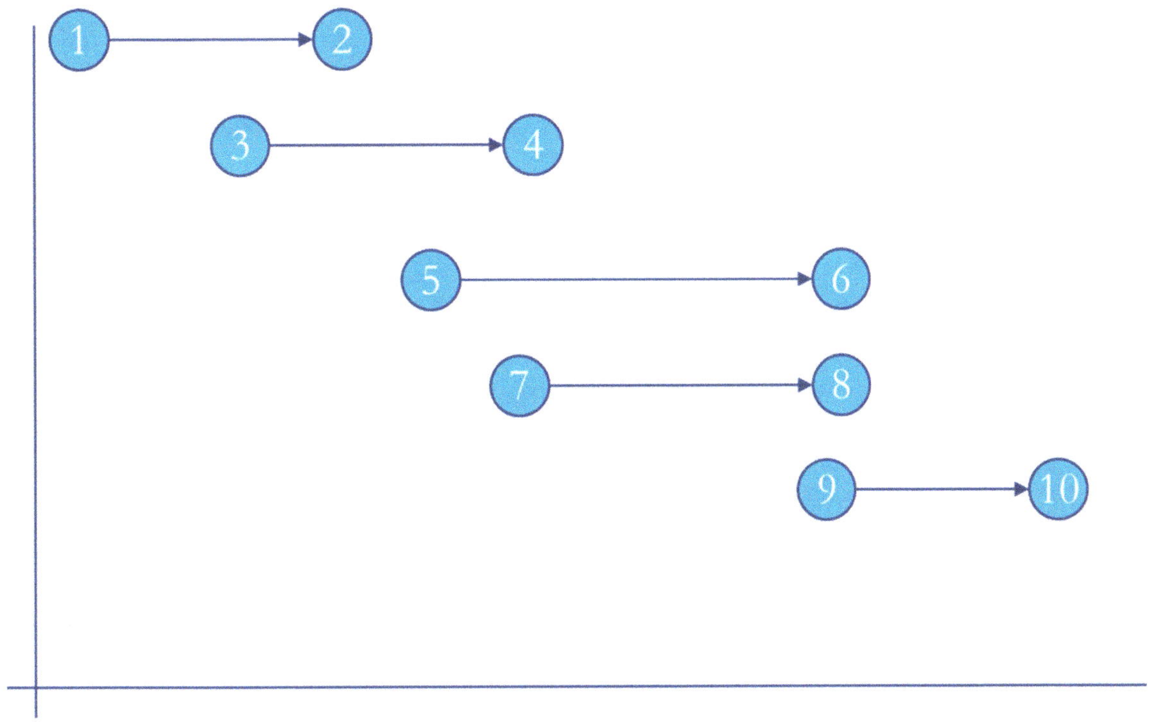

Figure 20: Events assigned to Gantt

We then assign relationships between the events representative of the relationships between the tasks on the Gantt chart.

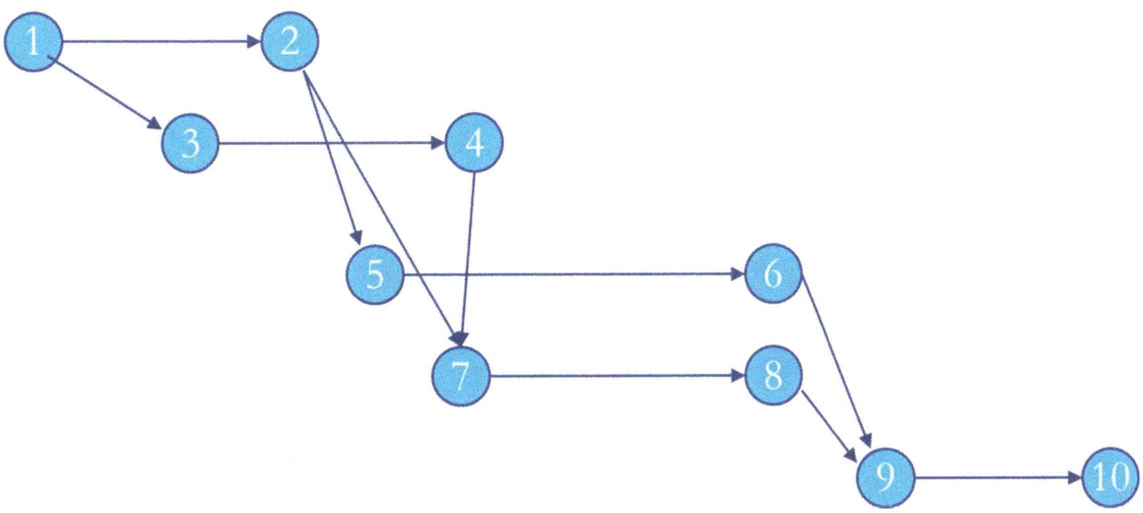

Figure 21: Network Diagram

Durations are added to the links and the critical path is identified.

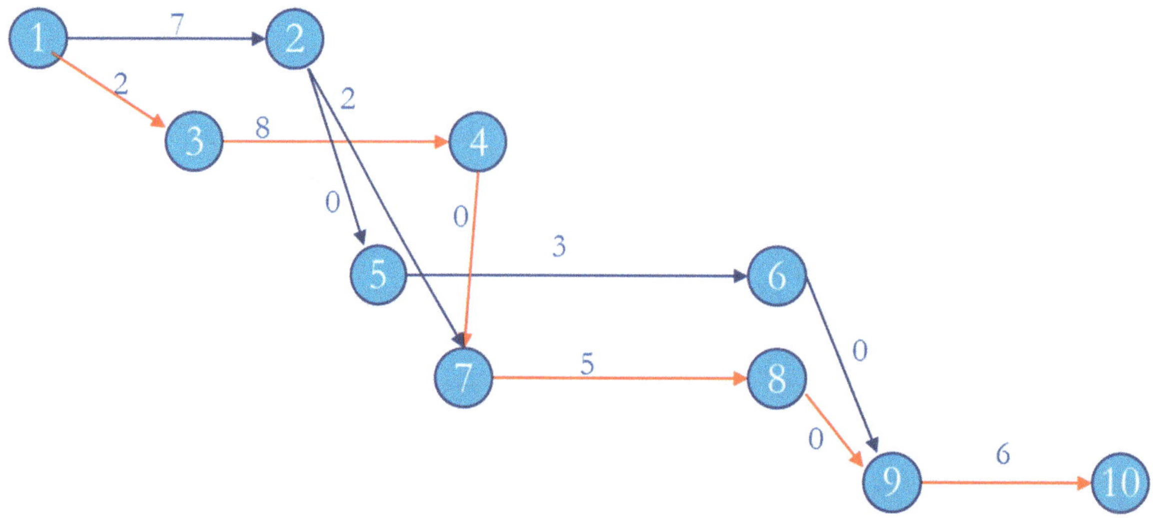

Figure 22: Network diagram with critical path

The Preferred Method of Schedule Generation

The method that is utilized in many large companies for the generation of an appropriate schedule for a project plan includes the development of an Integrated Master Plan (IMP) and an Integrated Master Schedule (IMS). This method is relatively straightforward.

There are three steps to completing the Integrated Master Plan.

1) Identify events
2) Identify accomplishments
3) Identify criteria

To complete an Integrated Master Schedule, you simply expand the task hierarchy to include all of the work packages necessary to accomplish the criteria, accomplishments and events. A simple software project is included as an example:

IMP Level	IMP ID	Task Name
Event	5.0	- Internal Integration & Pre-Qual Test Complete
Accomplishment	5.1	- GNC Software Builds Complete
Criteria	5.1.1	+ GNC Software Build 1 Complete
Criteria	5.1.2	+ GNC Software Build 2 Complete
Criteria	5.1.3	+ GNC Software Build 3 Complete

Figure 23: Integrated Master Plan

The addition of tasks will complete our Integrated Master Schedule in much the same way that we developed our Gantt chart. The completed Integrated Master Schedule is shown in Figure 24.

IMP Level	IMP ID	Task Name	CWBS	Duration	Early Start	Early Finish
Event	5.0	- Internal Integration & Pre-Qual Test Completed		189 d	9/1/09	5/21/10
Accomplishment	5.1	- GNC Software Builds Completed	1.1.6.5	189 d	9/1/09	5/21/10
Criteria	5.1.2	- GNC Software Build 1 Completed	1.1.6.5a	63 d	9/1/09	11/26/09
Task	5.1.2	Determine GNC SW Build 1 Requirements	1.1.6.5a	10 d	9/1/09	9/14/09
Task	5.1.2	Conduct GNC SW Build 1 Design	1.1.6.5a	18 d	9/15/09	10/8/09
Task	5.1.2	Code GNC SW Build 1	1.1.6.5a	25 d	10/9/09	11/12/09
Task	5.1.2	Conduct GNC SW Build 1 Unit Test	1.1.6.5a	10 d	11/13/09	11/26/09
Criteria	5.1.3	- GNC Software Build 2 Completed	1.1.6.5b	63 d	11/27/09	2/23/10
Task	5.1.3	Determine GNC SW Build 2 Requirements	1.1.6.5b	10 d	11/27/09	12/10/09
Task	5.1.3	Conduct GNC SW Build 2 Design	1.1.6.5b	18 d	12/11/09	1/5/10
Task	5.1.3	Code GNC SW Build 2	1.1.6.5b	25 d	1/6/10	2/9/10
Task	5.1.3	Conduct GNC SW Build 2 Unit Test	1.1.6.5b	10 d	2/10/10	2/23/10
Criteria	5.1.4	- GNC Software Build 3 Completed	1.1.6.5c	63 d	2/24/10	5/21/10
Task	5.1.4	Determine GNC SW Build 3 Requirements	1.1.6.5c	10 d	2/24/10	3/9/10
Task	5.1.4	Conduct GNC SW Build 3 Design	1.1.6.5c	18 d	3/10/10	4/2/10
Task	5.1.4	Code GNC SW Build 3	1.1.6.5c	25 d	4/5/10	5/7/10
Task	5.1.4	Conduct GNC SW Build 3 Unit Test	1.1.6.5c	10 d	5/10/10	5/21/10

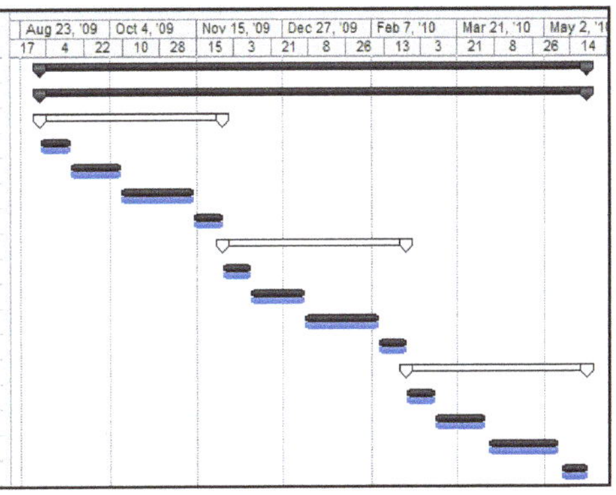

Figure 24: Integrated Master Schedule

This method is very straightforward and develops a highly detailed schedule that provides a coherent view of the project to be performed. As a final example in this section, we will provide the definition of a project schedule for a holiday dinner.

Project name: Holiday Dinner
Requirements:
Ten guests
Turkey with stuffing
Side dishes: potatoes, carrots, rolls, cranberries
Pre-meal wine

Dessert: apple and pumpkin pie

To define Events:
Holiday Meal

To define Accomplishments:
Holiday meal

- Invite guests
- Prepare for meal
- Prepare meal
- Serve meal
- Serve dessert
- Reset home

To define Criteria:
Holiday meal

- Invite guests
- Prepare for meal
    - Clean house
    - Prepare kitchen
    - Prepare dessert
- Prepare meal
    - Prepare turkey
    - Prepare side dishes
    - Set dining table
- Serve meal
    - Serve wine
    - Serve dinner
    - Clear dinner
- Serve dessert
    - Serve dessert
    - Clear dinner
- Reset home
    - Clean kitchen
    - Clean remainder of home that has been used

We then detail the task structure, creating the IMS (Integrated Master Schedule):

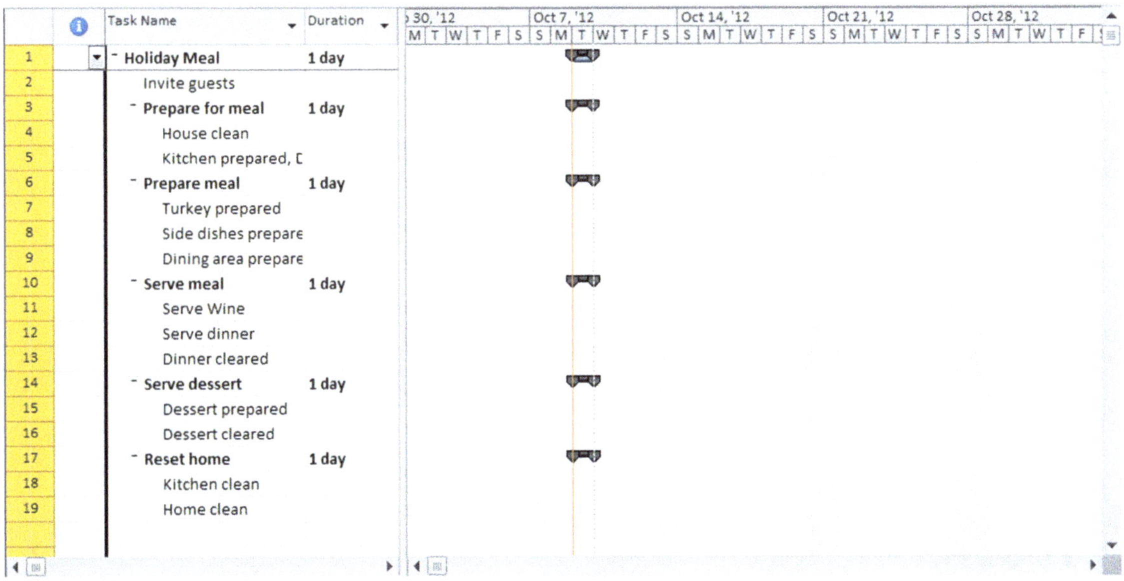

Figure 25: IMS

Add the task durations and link to tasks together:

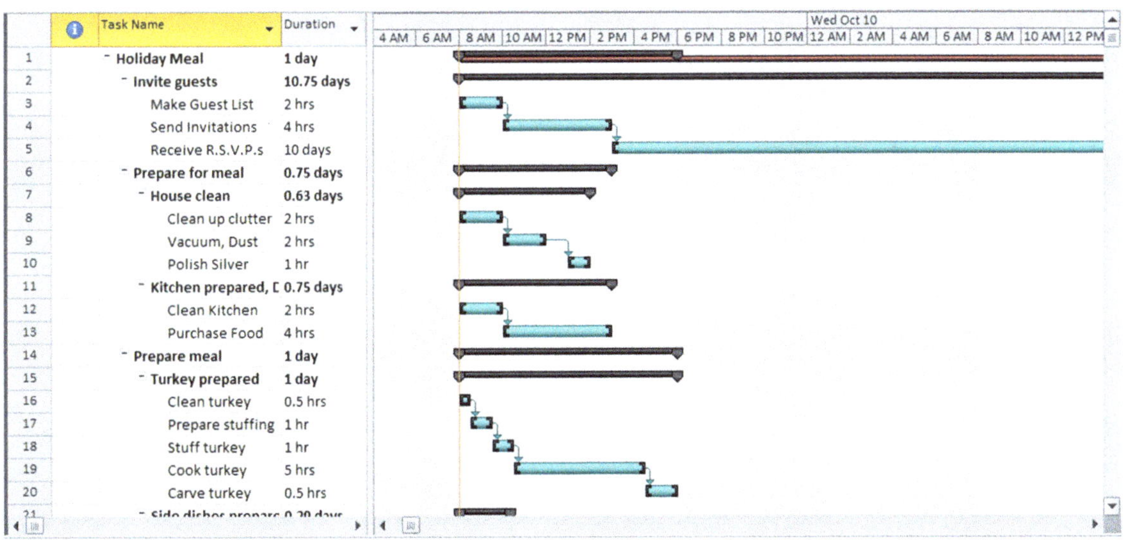

Figure 26: IMS with Links

Link the subprojects together:

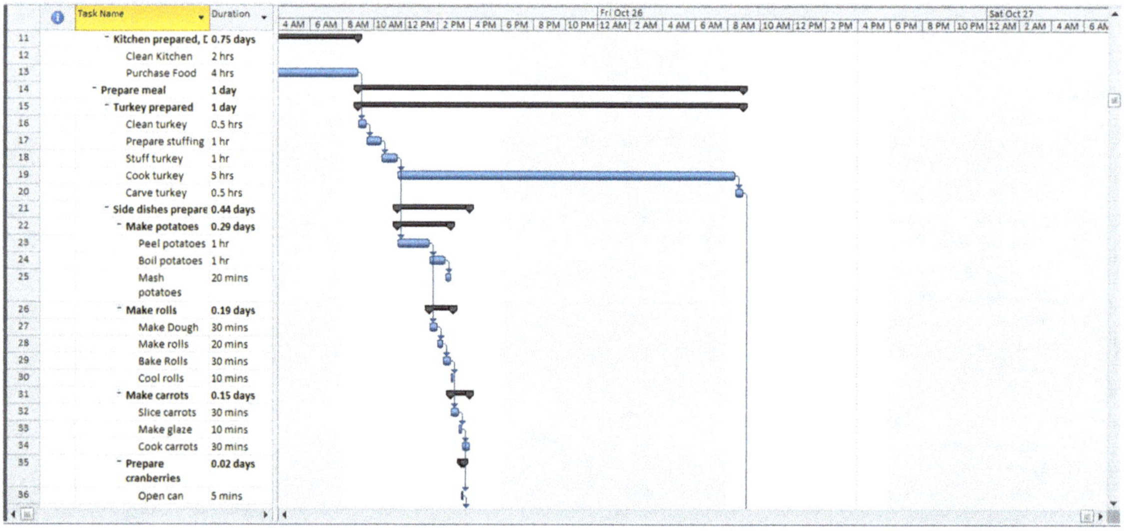

Figure 27: Linked IMS

Then assign resources and complete your project schedule.

ID	Task Name	Duration
1	**Holiday Meal**	**1 day**
2	**Invite guests**	**10.75 days**
3	Make Guest List	2 hrs
4	Send Invitations	4 hrs
5	Receive R.S.V.P.s	10 days
6	**Prepare for meal**	**1.38 days**
7	**House clean**	**0.63 days**
8	Clean up clutter	2 hrs
9	Vacuum, Dust	2 hrs
10	Polish Silver	1 hr
11	**Kitchen prepared,**	**0.75 days**
12	Clean Kitchen	2 hrs
13	Purchase Food	4 hrs
14	**Prepare meal**	**1 day**
15	**Turkey prepared**	**1 day**
16	Clean turkey	0.5 hrs
17	Prepare stuffing	1 hr
18	Stuff turkey	1 hr
19	Cook turkey	5 hrs
20	Carve turkey	0.5 hrs
21	**Side dishes prepar**	**0.44 days**
22	**Make potatoes**	**0.29 days**
23	Peel potatoes	1 hr
24	Boil potatoes	1 hr
25	Mash potatoes	20 mins
26	**Make rolls**	**0.19 days**
27	Make Dough	30 mins
28	Make rolls	20 mins
29	Bake Rolls	30 mins
30	Cool rolls	10 mins
31	**Make carrots**	**0.15 days**
32	Slice carrots	30 mins
33	Make glaze	10 mins
34	Cook carrots	30 mins
35	**Prepare cranberries**	**0.02 days**
36	Open can	5 mins
37	Slice cranberries	5 mins
38	**Dining area prepar**	**0.13 days**
39	Prepare linens	30 mins
40	Set table	30 mins
41	**Serve meal**	**0.38 days**
42	**Serve Wine**	**0.11 days**
43	Open wine	5 mins
44	Serve wine	15 mins
45	Open wine	5 mins
46	Serve wine	30 mins
47	**Serve dinner**	**0.17 days**
48	Invite guests to table	5 mins
49	Serve turkey	5 mins
50	Serve side dishes	5 mins
51	Serve wine	5 mins
52	Replenish dishes	1 hr
53	**Dinner cleared**	**0.04 days**
54	Remove plates	10 mins
55	Remove food	10 mins
56	**Serve dessert**	**0.06 days**
57	**Dessert prepared**	**0.04 days**
58	Take pie order	5 mins
59	Slice pie	10 mins
60	Serve pie	5 mins
61		10 mins
62	**Reset home**	**0.44 days**
63		1.5 hrs
64	Home clean	2 hrs

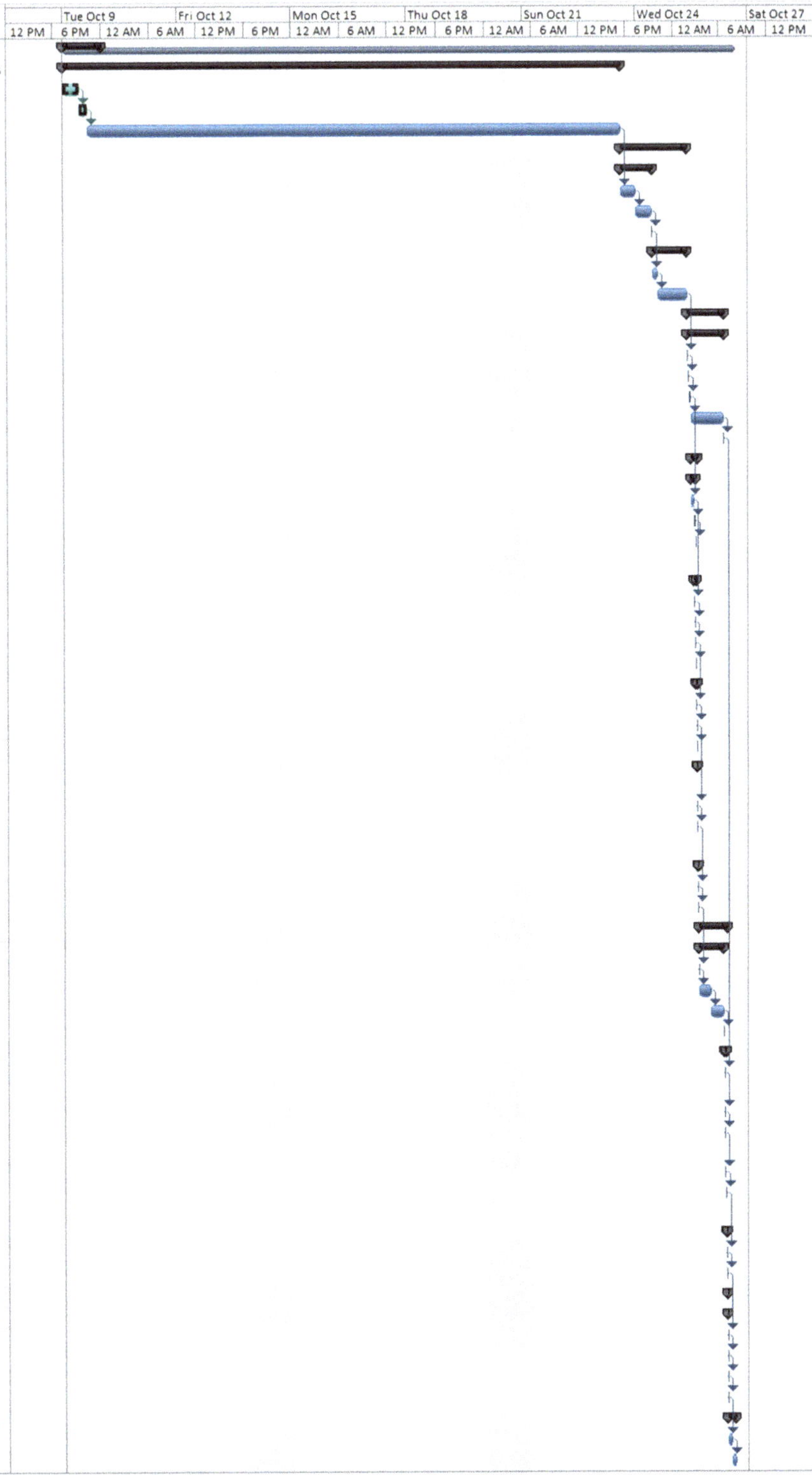

Figure 28: Completed Project Schedule (IMS)

Evaluating Risk

Risk assessment is critical to the success of the Project Manager. It is common for the average Project Manager to begin the plan with a high likelihood of failure. This is largely due to the planning of a project irrespective of other work occurring within the organization. Risk assessment allows us to define the potential challenges that will arise during the project execution phase and account for them in our planning.

To define risk, consider each task as a separate project. We can postulate that a given task repeated by a wide variety of people will result in a variety of outcomes. In this case, we are interested in the completion time of the task. We see by the following diagram that the resulting outcome of the task can be modeled by a normal distribution.

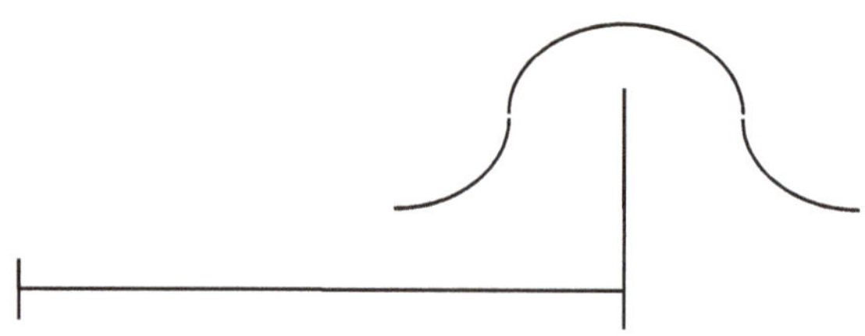

Figure 29: Task Completion Distribution

The issue of risk assessment is then related to defining the approximate shape of the distribution about the nominal task end. This shape will define for us the likelihood of completing the task as expected. Taken as a whole, all tasks within a project can be assessed in this manner, allowing a confidence interval to be calculated for the entirety of a project.

The information to define this distribution can be collected in a straightforward manner as shown in the following table.

Task #	Nominal Duration	Worst-Case Duration
1	4d	5d
2	1d	1d
3	5d	40d

4	6w	7w
5	10d	11d

As expected, the nominal duration is the intended schedule time to complete the work. The worst-case duration is defined to assess the approximate width of the standard deviation. It is surprising to observe that several of the tasks show a distribution width that exceeds the task length. This is an insight into the psyche of the project planner, as he or she does not desire to communicate that the task schedule is thirty-five days plus or minus one year. This creates a significant credibility issue for the planner, and is therefore mentally suppressed.

We desire to communicate a schedule with realistic risks associated, and therefore a worst-case scenario is chosen as the appropriate method, bypassing this blind spot in the planner's viewpoint. As we observe from the above table, the risk of certain tasks is quite low and predictable. Other tasks show a great deal of risk, such as task three. These of course are the tasks requiring the attention of the Project Manager, as the greatest variability in response will be observed in this area.

For example, consider a project made up of the following tasks:

Task #	Nominal Duration	Worst Case Duration
1	10d	15d
2	10d	20d
3	10d	10d
4	10d	13d

The project at the outset shows expected times of completion:

$T_{nominal}$ = 10d + 10d + 10d +10d = 40 d

$T_{worst\ case}$ = 15d + 20d + 10d + 13d = 58 d

This provides us with an idea of the standard deviation, assuming the worst case shows a 3-sigma distribution.

$\sigma = 6$ d

Therefore, if we are asked if we are confident that we will complete the project in forty-six days we statistically predict that this encompasses a nominal value of 40 d plus a one-sigma area of the distribution. This corresponds to a confidence interval as follows:

$$C.I._{46days} = 67\%$$

In the event twenty-six days have passed (about half the management allowed time) and the first two tasks are completed, we can recalculate the confidence interval.

$$T_{nominal} = 10d + 10d = 20\ d$$

$$T_{worst\ case} = 10d + 13d = 23\ d$$

$$\sigma = 1\ d$$

We must consider our confidence now of completing the remaining tasks within twenty-two days:

$$C.I._{22days} = 20\ days + 2\ sigma = 93\%$$

Therefore, although we are approximately half completed with the project and half completed with the tasks, our confidence interval has increased due to the lower risk tasks still remaining to be completed. If we are able to shift the higher risk tasks further forward into the project, or if we are able to focus directly upon these tasks, we can have the greatest impact on the overall project delivery.

Imagine that in the above example, the tasks are identical but the first two tasks are completed within 25 days.

Task #	Nominal Duration	Worst-Case Duration
1	10d	15d
2	10d	20d
3	10d	10d
4	10d	13d

$T_{nominal} = 10d + 10d = 20\ d$

$T_{worst\ case} = 10d + 13d = 23\ d$

$\sigma = 1\ d$

We must consider our confidence now of completing the task within 23 days:

$C.I._{23days} = 20\ days + 3\ sigma = 99.7\%$

Therefore it is very important for the Project Manager to establish the risks within the project appropriately to define the remaining risks as the project proceeds. Additionally, it should be noted that based on this model, all projects not including risk calculations are currently planned with a confidence interval of only fifty percent.

If we feel confident that our risk calculations are representative of the expected variability in our project to be executed, we incorporate the risk (additional project duration known as management reserve) into our project schedule. This addition of time to reflect risk is used in order to define a higher probability of finishing the project on-time.

In other words, if we have a project scheduled for 40 days of work, and we calculate that the worst case indicates a 58-day duration, we may choose to accept a 93% probability of completion and plan our project for a 52-day duration. Thus we are expecting that our 40 days of work will occur (float) somewhere within those 52 calendar days, as other work and interruptions will interfere with the completion of our project.

Once we have completed the Integrated Master Schedule, we have completed the second task in our project planning challenge, namely we have defined 'how long it will take'. Next we move on to developing the budget.

Budget Definition

To define an appropriate budget for our project, we must first make sure that our Integrated Master Schedule makes sense. To accomplish this, we will meet with the Functional Managers to ensure the schedule we have generated has realistic task identification and durations that will produce a reasonable expectation of being able to complete the project as planned. We meet with Functional Managers because they are the subject matter experts for the areas of interest, and it is very likely they will help us to produce a more accurate and representative schedule. What's more, since the Functional Managers will assign their personnel to this project, this is our first step at socialization of our upcoming project plan.

For each of the work package tasks within our Integrated Master Schedule, we will assign costs as appropriate for that task. The resources that are utilized to accomplish that task and their associated billing rates typically represent these costs. These billing rates are multiplied by the duration of the tasks to generate the actual cost of those resources accomplishing that task.

Additionally, we will include other costs associated with those work packages such as:

Indirect labor hours
Fixed costs
Materials
Purchased parts
Subcontracts
Freight
Travel costs
Other miscellaneous costs (10-15% add-on)

These can be incorporated into a simple project budget worksheet, such as included here for reference.

Cost Estimating Worksheet

Project Title:			Project Manager:			
Activity	Schedule	Budget				
No.: Name/Description	Hours	Labor hr x rate	Material	Other	Total	Resource Manager
Prepared by:			Date:			

As the budget is completed, we have accomplished what we originally set out to accomplish.
We have defined:
What will we do?
How long will it take?
How much will it cost?

Our only remaining task is to complete the Project Charter. The charter document is a single document that incorporates all of the work accomplished to date on the project plan.

The Project Charter Document is used in a similar manner as a contract within the organization. The information is filled out at the completion of the project planning process, and defines exactly what will be accomplished on the project, what will not be accomplished on the project, the expected costs and the expected timing of the project. Space is provided to gain approval signatures from stakeholders on the project and the intent of the project charter is to use the document both to open and close the project.

An example Project Charter Document is included to provide some insight into the technical information required. We have already defined all the information required to complete the document, but the charter acts as a single repository for the information.

Project Charter

Prepared By:	
Version:	
Date:	

Sponsor:	
Project Manager:	

Approval Signatures

Name ____

Title ______________________________ Date ______________

Name ____

Title ______________________________ Date ______________

Final Customer Sign Off*

Name ____

____

* Should be completed as a final signoff to the project.

Title ______________________________ Date ______________

Name ____

Title ______________________________ Date ______________

Key Stakeholders			
Project Sponsor:		Phone:	
Division:		Department:	
Project Manager:		Phone:	
Division:		Department:	
Colleges / Departments Affected:			

Summary:	How does this project contribute to the overall Vision of the Organization? With which strategy does this project align?

Problem/ Opportunity	What is the problem or opportunity that this project poses to correct or take advantage? (Not the project benefits)

Benefits	Identify the Tangible and Intangible benefits that will be derived from this project implementation.
Tangible: Intangible:	

Critical Success Factors	Critical success factors are those business factors that should be accomplished in order to consider this project a success.
1.	
2.	
3.	
4.	

In Scope	Describe the items that are in scope for the successful completion of this project. Any items not listed are not understood as being in scope.

Out of Scope	Describe the items that are out of scope for the successful completion of this project. Any items not listed are not understood as being out of scope.

Risks / Opportunities	What risks or opportunities are associated with completing this project, or not completing this project? What risks or opportunities can be identified when considering resources needed to complete this project?

Costs	What are the costs associated with this project *(Cost are usually estimated with a confidence rating at the beginning and continue to firm up during the planning process.)*

Costs:				
Hardware	$ -		Consulting	$ -
Software	$ -		Internal Labor	$ -
Support Contract	$ -			
Other	$ -			
Expense Total	$ -		**Labor Total**	$ -
	Total Cost:	$ -		
Initial Cost	$ -			
Monthly Cost of Ownership	$ -			
Monthly Benefit	$ -			
Cummulative ROI	#DIV/0!			
**Based on a three year rate of return*				

ROI is calculated as 1-[{Total Cost + (Cost of Ownership * 36)} / (Monthly Benefit * 36)]

Interdependencies	In order to complete this project, does it require any other projects or tasks outside of the Project Team's control to complete in order to be successful?

Milestones	In this section the user can identify the top four milestones and their proposed completion dates.	
Milestone	Description	Completion Date
1.		
2.		
3.		
4.		

Technology	What technology will be used to deliver the solution?

Team Composition	How will the Project Team be composed?

Stakeholders	List all of the individuals, departments or users who will be impacted by the outcome of the project.

Resources	What resources are needed to complete this project in scope and on time? Resources include people, skills and equipment. Include an estimate to the time a resource is needed.

Assumptions	What assumptions are being made about the resources, budget, schedule, etc.?

Constraints	List major constraints that are on the project team and the completion of this project. Be sure to include milestone dates, resource availability and budget.

To fully understand the Project Management lifecycle, there are additional phases of the project that must be understood. These include Project Control, Project Execution and Project Close-out. Many textbooks exist regarding these phases of the project, and the finer points of several of these areas go beyond the scope of this text.

Phase 3: Project Control

Details of Project Control are included in the Earned Value Management section of this text. In that section, a method is presented that allows appropriate status definition and reporting on the project during the execution phase. That chapter outlines the details that are adequate for the majority of users to understand how to track and report on a project, and discusses the benefits of making reasonable predictions regarding the outcome of the project during the project execution.

Phase 4: Project Execution

Project Execution is probably the most convoluted and challenging phase of any project endeavor. This is the portion of the project in which the project leader must rely upon his or her ability to provide an engaging project plan that will result in the employees investing their efforts in the successful completion of the plan. This covers a wide variety of topics, including individual motivation, communications, social styles and personality styles, group behavior, sociology, psychology and anthropology. In general, technical professionals gain these skills through application of the project techniques previously mentioned, and often must learn by observation in the real world. It is of great benefit to the project leader, however, to investigate the wide variety of publications available to assist in this area of project execution.

The Role of Complexity and Chaos in Project Management

Discussions with individuals involved in the Project Management function will often produce the same list of Project Management problems. These problems will typically include items such as the following:

- Creeping Scope (Performance)
- Unrealistic Expectations
- Lack of Planning
- Lack of Control
- Poor Tracking
- Anomalies occurring during the Project
- Chaos throughout the project
- Others ...

Given the opportunity, the list could be extended to several pages. What is interesting, however, is the common thread in all the problems listed; that is, the appearance of Chaos within the structure of the project. This Chaos is frequently credited with the translation of a project with an exquisite project plan into a jumbled mess that is virtually unmanageable.

Unfortunately, it is not possible to tell you that your projects will be entirely manageable. In fact, what is closer to the truth is the assertion that all projects exist within a Chaotic environment, and will experience Chaotic periods. That's the bad news. The good news is that Chaos is understandable and manageable on a system level. You can manage a project in a Chaotic environment if you are willing to allow the project to act in a natural manner.

What is Chaos and Complexity?

Chaos is a relatively new branch of mathematics that has received a great deal of attention in recent years. One of the more straightforward explanations of the phenomena may be found in Bak (1996). In this profoundly named text, Bak describes from an empirical perspective how the realm of Chaos exists all around us. Bak notes that Chaotic models can be used to describe earthquake patterns, the use of English words in this textbook, and the pattern of behavior of the U.S. stock market.

Some have taken issue with Bak's work, stating that his model is nothing more than an empirical description of complex behavior, and is therefore of little use to the practitioner. One need only consider the fact that the majority of our natural sciences knowledge was developed by similar means to recognize the potential fallacy of this argument.

Bak proposes a model of Chaos that exists in the physical domain between order and randomness. This is shown in Figure 30.

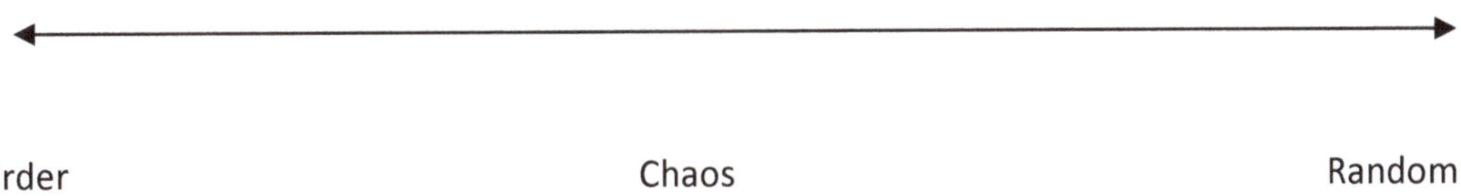

Figure 30: Chaos in the Physical World

Obviously there is some hope, as Chaos is not a random process. It is clear that Chaos is more difficult to understand than an ordered model. In fact, Chaos is a phenomenon that exhibits the behavior of a complex system. The system is so complex that we are unsure of all the constructs contributing to the behavior.

This type of model reflects the real world of Project Management. In short, Project Management is impacted by a wide variety of sources within an organization, many of which are beyond our reach or awareness. Thus, the study of Chaos will enable us to better understand the overall system of project management.

A Simple Chaotic Model: The Sandpile Model

It is beyond the scope of this text to describe in detail how Chaos and complexity function. It is sufficient for the understanding of the Project Manager to investigate the behavior of two simple Chaotic models. The first of these models is the Sandpile model. The second model is the Metered Traffic model.

The Sandpile is described in detail by Bak (1999, *How Nature Works*. Springer. New York) and can be implemented with a simple deck of cards. An exercise of this type will enable the Project Manager to very quickly grasp the behavior of a Chaotic system. First, we must define some of the behaviors of a complex system:

They operate under a set of simple rules
They self-organize into a state of criticality
Systems in a state of high criticality are subject to disturbances

To model a simple Sandpile, place nine playing cards on a table in a three by three matrix as shown in Figure 31.

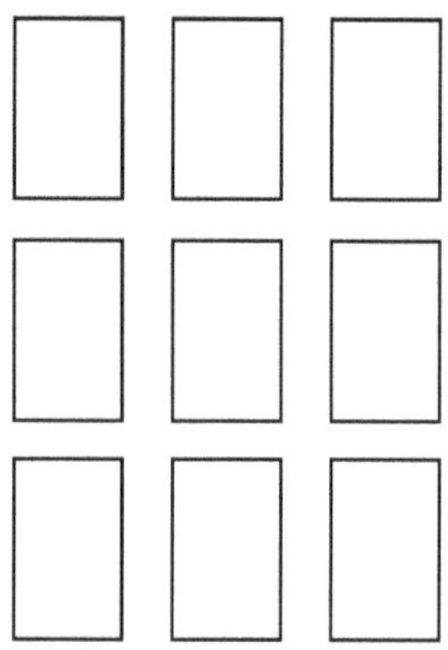

Figure 31: Sandpile Matrix

The 3X3 matrix will represent a Sandpile, and the addition of single cards to the system will represent a single grain of sand being dropped onto the Sandpile. As might be expected, when enough grains of sand are dropped into the matrix, some avalanching will occur. This effect is described with a simple set of operational rules for the simulation. These rules are as follows:

1. Introduce a card randomly to the matrix.
2. If there are four cards in any pile, these cards avalanche into the N, S, E and W piles adjacent to the original pile. This is shown for the center pile in Figure 32.
3. If a pile is on the edge of the matrix, cards moving out of the matrix are considered to have exited the system.
4. Record how many cards moved during the process.
5. Continue introducing cards into the matrix until you have exhausted the original deck.

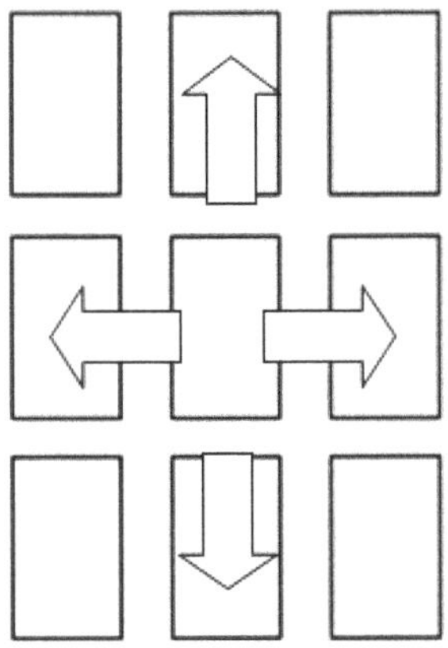

Figure 32: Avalanche from the Center Pile

It becomes very clear to the individual experiencing this simulation that a large number of cards can move with the introduction of a single card. This typically happens only after a number of cards have been introduced into the matrix, thereby increasing the criticality of the system. As criticality of the system reaches a critical mass, avalanches become the norm. These catastrophes or anomalies are therefore present in all systems, and the frequency of their occurrence is dependent upon the critical mass of the system. Remember those anomalies that occurred on your last project? They are a part of every project you will ever manage, and we must discover how to minimize the frequency of occurrence of those anomalies.

The 3X3 Sandpile model is very small, and therefore easily implemented manually. It can be thought to represent the behavior of nine simultaneous projects operating within a single organization. Upon investigation we discover that the edge effect experienced by the system is large, caused by the fact that eight of the nine cards form an edge of the matrix. Therefore a larger matrix might be more informative. However, a larger system rapidly becomes unwieldy for the user, and is best modeled on a computer.

It is clear that many grains of sand must be introduced into the system before any avalanches occur. As the critical mass increases, the avalanches become more frequent, until eventually they become the norm. (Note that the numbers recorded in Figure 33 represent the number of avalanches, representing the movement of four grains of sand.)

What is of interest to the Project Manager is a perspective on the patterns occurring in a Chaotic system such as this. The frequency distribution of a 25X25 matrix is compiled in Figure 33.

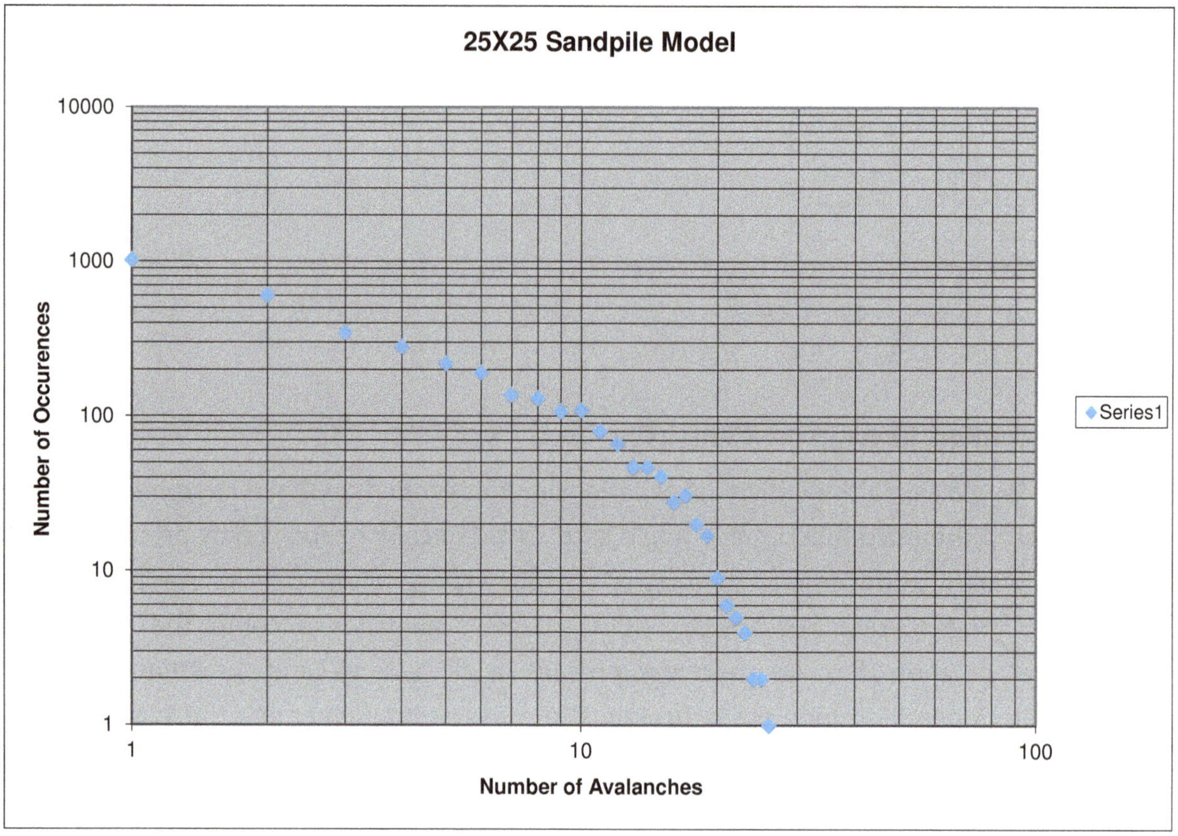

Figure 33: Frequency Distribution

Thus the behavior of a Chaotic system follows a power law. Note that the curving downward of the line is caused by both edge effect and sample size. Larger data sets do not exhibit this phenomenon.

The project exists within a Chaotic system. There are numerous other projects within the system, all bumping into each other. The single project system will exhibit many small magnitude catastrophes throughout its duration, as well as fewer large catastrophes. These are the anomalies discussed earlier. They will exist in every project, and we must manage the project to account for their existence.

A Model of Chaos: Metered Traffic

A second model of Chaos is demonstrated in the behavior of metered traffic. What is noteworthy in this example is that a Chaotic problem is presented with a solution for handling this problem. The method of solution is applicable for the Project Manager as well.

Metropolitan areas frequently experience traffic problems due to excessive volume. One method utilized to alleviate this problem is metering the flow of traffic into the system. This is typically accomplished on entrance ramps to freeway systems. Some have argued that this simply transfers the problem from freeways to surface streets. However, as we shall see, this is not the case.

The traffic system operates under simple set of rules. The traffic flow is analyzed, the difference between the actual traffic flow and the maximum possible flow is calculated, and the rate at which cars are metered into the system is set. The design of the system is focused on the maximization of traffic flow through the system. Obviously this system has all the elements necessary for Chaotic behavior.

The system was designed to maximize the flow of traffic, which is essentially maximizing the critical mass of the system. This would imply that the system would be more susceptible to catastrophe brought about by a small disturbance. Remember that the small disturbance in the Sandpile model was the introduction of a single grain of sand to the system. In the traffic case, the disturbances may be sourced from those elements already present in the system as well as those entering the system. As a tangible representation of an avalanche, imagine what occurs when a driver of one of those cars applies the brakes. The drivers immediately following also apply the brakes, and the disturbance avalanches throughout the system. In this case, a small tap on the brakes might impact ten thousand cars.

The solution presented by the traffic managers was to reduce the critical mass of the system. Instead of metering cars at a maximum capacity level, they instead chose to meter cars at a seventy-five percent level. This results in a buffer zone around each car, so that when brakes are applied it affects far fewer vehicles, perhaps ten cars. Thus the reduction in criticality of the system results in smaller, more manageable catastrophes. There is a counterintuitive conclusion to be drawn from this. *If cars are metered at a less than maximum capacity, more cars get through*. Although waiting time to enter the freeway is increased, overall trip time is reduced in the general case. Consistency is increased, and therefore the quality of the system is increased.

This conclusion was extended to the manufacturing factory floor, where Industrial Engineers reduced the rate of the bottleneck machine in the plant from maximum capacity down to seventy-five percent capacity. Again the result was counterintuitive. They were able to make more parts. *By reducing capacity, they were able to make more parts.* Chaos plays a very significant role in our lives. Lack of planning for Chaotic behavior can be a very costly oversight.

As a last note, one can discover many Chaotic models about us in the world. These models typically exhibit behavior that is more complex than is easily understood on a detailed level. The system level is more easily understood, such as in the case of the Windows operating system. The behavior is Chaotic, but yields an appropriate complex behavior for the system.

What about the Project Manager?

Dr. Fred Zimmerman (1997) most succinctly stated the role of the Project Manager when he noted that the Project Manager must embrace the fact that as a Project Manager, he or she will always be in trouble with someone, the question is just with whom. Likewise, Chaos exists on every project. Acceptance of this fact enables the Project Manager to accept the Chaotic behavior of the system as a necessary consequence of operating in an imperfect environment.

If we could insulate every project, so that it would never be impacted by other projects within the organization, we could significantly reduce or eliminate Chaos from our list of concerns. However, as Peters (1988) has pointed out, all white-collar work is essentially project oriented. This would imply that all Project Managers are faced with the likely occurrence of Chaos within their daily activities.

What can be done to manage a project within a Chaotic environment? The project manager must recognize that excess resources must be committed to the project in terms of management reserve, to account for the complex behavior of the organization. As stated in the traffic example, it was discovered that a committed reduction to 75% of the resources resulted in superior traffic flow. Thus the Project Manager must schedule the resources on the project to not exceed 75% utilization.

In project terms, this means that three days of effort might be required to complete a specific task. The duration utilized in the budget will reflect a three-day effort to accomplish the task. The schedule, however, will reflect a four-day completion cycle. This communicates that three days of work will be accomplished, but it will take four days to finish the work. Why is this the case?

Phone calls
Email
Voice mail
Conversations
Interruptions
Other projects

Other people
Anomalies
Others

The counterintuitive conclusion to be drawn from this discussion is as follows:

If you schedule projects for seventy-five percent capacity, you will get more work accomplished.

Project Managers will increase their success rate by reducing the rate at which they schedule projects. In many cases, organizations schedule personnel on projects at one hundred percent utilization, or worse yet one hundred twenty percent utilization. This is accomplished through a company culture that demands mandatory overtime of the employees. Project Managers must plan for Chaotic behavior on their projects by reducing the utilization of the resources to a more realistic level. Is seventy-five percent an appropriate level to compensate for these effects? That is for the Project Manager to determine in his or her specific organization and application.

It is clear that Chaos can be managed to account for all those things that occur during a project that cannot be predicted or incorporated into the project plan. It is time for Project Managers to become forward thinking, and understand that those anomalies that caused problems on their last projects are ever present in the system and will appear on their future projects.

Phase 5: Project Close-out

Details of Project Close-out are included in the section of this text dedicated to business rhythm and reporting. The final phase of the project is a critical one, and a checklist is provided that allows a practitioner to review the current state of his or her completed project and compare it to the original plan, allowing some prediction of future performance on projects based upon the accuracy of those estimates and execution. For further development, the reader is urged to consult one of the many Project Management textbooks available.

Chapter 4: Program Management Culture

Program Management is an area of significant challenges to most technical organizations. The Program Manager is responsible for the entire lifecycle of a product family or program within the organization. This program will likely be made up of many different phases, which include concept refinements, technology development, systems development and demonstration, and production and deployment.

These responsibility areas extend far beyond the typical responsibilities of a Project Manager. In fact, many of these responsibilities extend well beyond the responsibilities of an engineering department. Therefore, the Program Manager must have a long-term viewpoint, and recognize their role is one steering the development of families of products or processes. This requires them to develop an ability to forecast and predict the behaviors of the marketplace, with enough forethought to allow the subsequent development of various systems and products to take place prior to a shift in the marketplace.

Additionally, the Program Manager must also deal with many departments that extend well beyond the engineering function. This will typically include finance, accounting, marketing, sales, production, packaging, shipping, quality, and many others.

It is clear that the program management function operates at a much higher level than that of the Project Manager. Although many of the skills are identical, a typical program will incorporate many projects under its umbrella. Therefore, the Program Manager is a manager of Project Managers in general. This requires a different set of skills than that discussed in the project-management domain, and also requires a different mindset in the execution of the program. These differences will be discussed in detail.

Programs and Projects

We must understand the distinction between programs and projects. Projects are typically relatively small in scale, with concrete objectives and finite timelines. Programs on the other hand might extend for many years and at the outset it might be quite unclear what is expected of the outcome of the program. Therefore, programs are much more nebulous and vague than are projects.

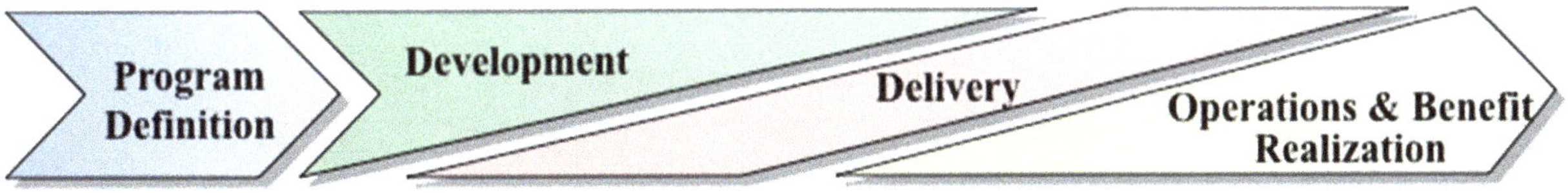

Figure 34: The Program Life-Cycle

Typically the program will spawn many projects in order to complete its diverse objectives. Many of these projects will reside within the Engineering domain, and many will reside without. It is the role of the Program Manager to integrate these very disparate projects and perspectives in order to achieve a harmonious whole.

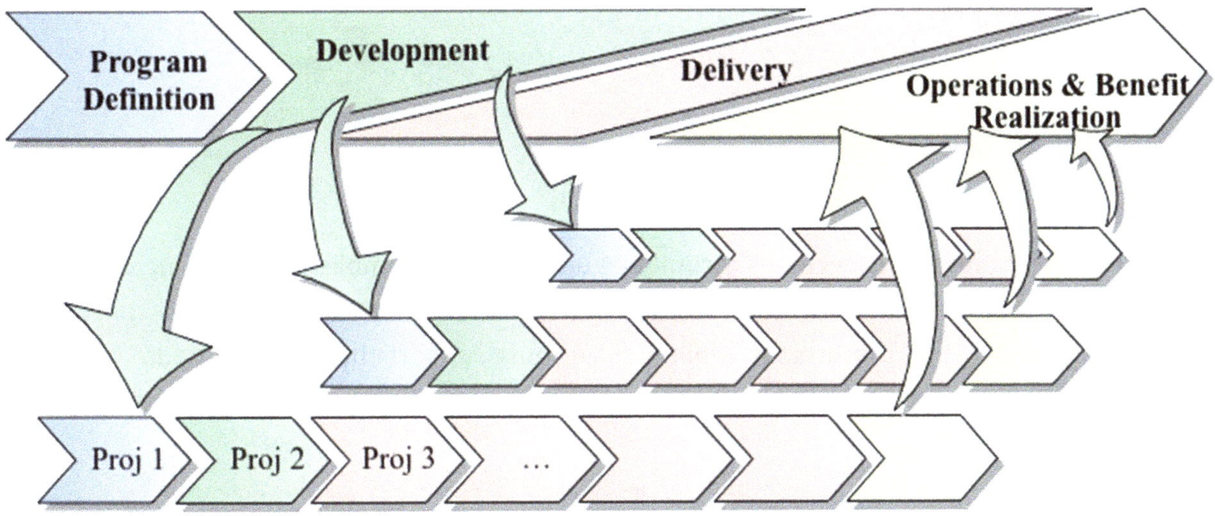

Figure 35: Programs spawn many projects

Programs and projects then obviously have very different lifecycles and durations. Although the skills to accomplish them are very similar, the outlook and perspective required of each is very different. The Program Manager is responsible for the long-term viability of the product and organization, while the Project Manager is responsible for the short-term completion of specific objectives as shown in Figure 35.

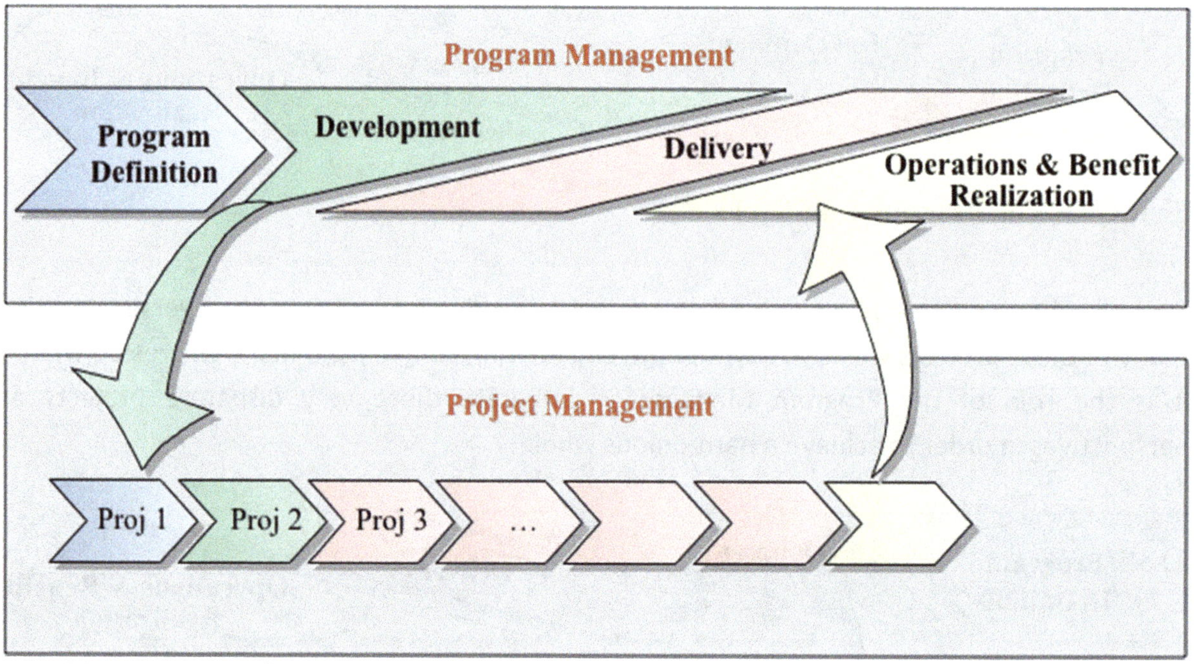

Figure 36: Program and Project Management

As an Engineering leader, it becomes incumbent upon us to accomplish Engineering objectives in a practical manner, meaning that we will plan and execute our projects as previously discussed. In addition to these responsibilities, we must support the product roadmap, which will clarify for us those products that are on our development horizon based upon the insights of the Program Manager.

Due to the significant complexity of the program management function, the engineering leader must also recognize that their responsibility extends into any development that might alleviate the responsibility of overseeing the projects on the part of the Program Manager (see Figure 36.) Thus a systematic and robust reporting system is necessary at the project level in order to support the program objectives. This is another compelling reason why the projects must be planned in detail and a realistic status must be defined and reported utilizing a business rhythm as seen in Figure 37.

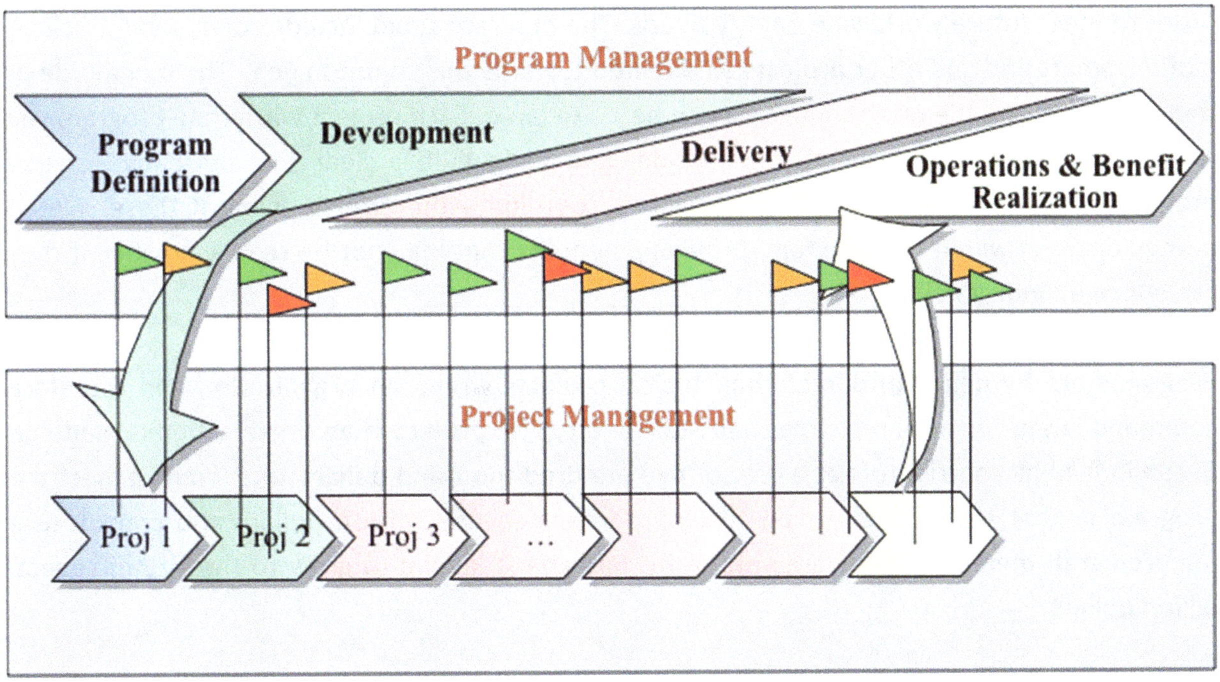

Figure 37: Project Status is Reported Regularly in Support of Programs

Program Management differs from Project Management primarily in scope and scale. There is a greater level of responsibility for the Program Manager, which requires a high level of business knowledge and capability. The magnitude of the differences between Project Managers and Program Managers is quite large, as the Program Managers must be very astute in political and diplomatic skill areas.

Program Management is clearly a step up from the Project Management domain. It is typically of more importance to the organization, deals with more complex efforts, and commands a higher degree of respect within the organizational hierarchy. Programs are strategic and typically include two or more interdependent projects that share a common and significant goal. The major goal of a program differs from a project as it is focused on strategic alignment and business success. Therefore, the focus of the program is about competitive advantage and sales. Add to that fact that Programs include multiple projects; it is by definition of a higher level of complexity. Program Managers also are required to perform functions that go well beyond those of a typical Project Manager; such as defining the acquisition process and establishing the resources, funds management, contracting and logistics, to name a few.

The scope of the program may cover many different types of projects from many different divisions or organizations, each of which are in some manner sharing the same goal. For example, a large company may develop a program to develop alternative energy methods in order to meet future worldwide energy needs. The program could include solar, wind, nuclear, biofuel, smart grid and other projects all directed towards this common goal. The magnitude of these projects and the program overall can be quite large. Each project within the program will retain its own goals, specification, schedule, and budget. It is clear the alternative energy program is designed to act as an umbrella to coordinate the efforts of all of these related energy projects within the program, providing benefits that may not be readily obvious if they are managed individually.

Programs are by definition larger than typical projects within an organization and therefore command larger financial resources and staff. If a typical project in an organization is made up of ten individual contributors at a cost of five hundred thousand dollars, one would expect the program in that same organization to cost perhaps several million dollars and constitute a hundred staff members. Very large programs have been known to grow to the size of several billion dollars.

Program Managers must bring skills in the areas of leadership and process development to the program. The leadership skills are similar to those of Project Managers when scaled up to the program level. In Program Management, leadership skills will focus on several specific areas of responsibility:

The development of a program vision that will be used to steer and engage project teams and their efforts.

The development of a program culture that establishes a trustworthy and professional work environment.

Motivation of the program contributors to reach a well-understood goal.

Collaboration across organizational boundaries in order to share resources.

The key challenge for any Program Manager is to establish a balance between the leadership of the program and the processes necessary for program success. The major focus and responsibility of the Program Manager will be the horizontal integration of all related efforts associated with the program even though the projects within the program can be quite different in scale, scope and size, with appropriate integrated processes that are consistently applied across the organization. These activities vary by organization but include planning methodologies, communication plans, metrics and their generation, resource allocations and risk management methods.

Establishing program goals and priorities are major drivers in the integration and decision-making of any program. All efforts should be aligned and directed towards meeting the needs of a specific business strategy. The Program Manager assumes full accountability for the success of the program and achievement of business goals, and will also more than likely share in financial success. Successful Program Managers will maintain a clear connection between organizational strategy and program execution.

A major challenge for the Program Manager is to integrate and manage the processes within the program (and existing organization), to enable the achievement of program goals and objectives. A major portion of the Program Manager's time is therefore taken up in the creation and development of these processes. Coordination of the complex interdependencies of a program can be a big challenge, and yet management of all of these moving parts is necessary in order to achieve program goals. The demanding amount of coordination required in a complex environment is why the Program Manager is typically a highly skilled and capable individual.

A relatively simple structure can be used to describe the varied the responsibilities throughout the life of a typical program. Without a clear definition of roles and responsibilities, often the vagueness of the requirements for each team member can result in a highly convoluted program effort. One technique used to describe these various responsibilities is a Swim Lane Diagram. An example of this is shown in Figure 38.

As can be seen in the diagram, the existing products are defined at the top level. The needed products and capabilities are defined at the program management level, which in turn defines when requirements must be established. The Systems Engineering effort then must be started and so on throughout the typical project lifecycle. In this example, a new product development program consists of approximately twenty individual projects spanning multiple functional areas.

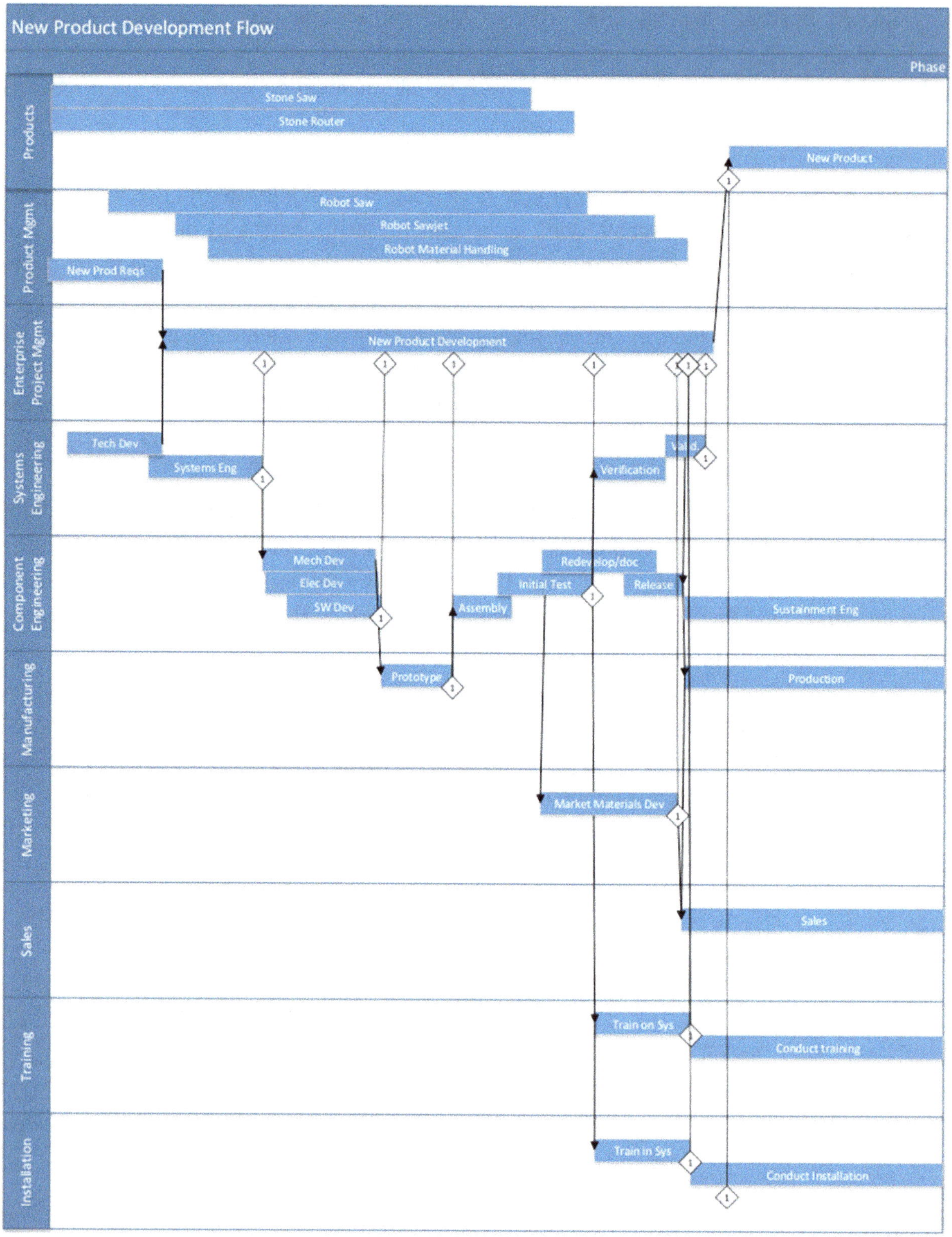

Figure 38: Program Management Swim Lanes

Program:

A program is a group of projects managed in a coordinated way to obtain benefits not available when managing them individually.

Many programs will include elements of ongoing operations. For example: The ABC vehicle program includes both the project (or projects) that will design and develop the vehicle as well as the ongoing manufacturing and sustainment projects associated with that vehicle in the field.

Programs may also involve a series of repetitive or cyclical undertakings. For example, major construction companies often run cyclical construction programs; an ongoing operation involving interrelated projects. Many nonprofit organizations will run a fund-raising program on a regular basis to sustain their ongoing efforts to obtain financial support for the organization.

Project Portfolio:

A portfolio of projects is a collection of projects owned or managed by the same organization that are only indirectly related. These projects may not be directly related with regard to inputs and outputs, meaning one project does not necessarily feed into another. Portfolios are typically chosen with a diversity of projects, some of which may be directly correlated and related while some may not appear to be related in any manner. Typically organizations will construct a portfolio of projects utilizing an aggregated project plan made up of several different types of projects. The aggregated project plan will contain some projects that are focused on advanced research and development. Secondly, the organization may focus on new products or processes that result in breakthrough projects. Platform projects focus on advancements in product families within the organization. Derivatives will focus on enhancements to existing products and processes.

Advanced R&D projects are typically identified as innovations in technology development that will be necessary prior to embarking on a commercially viable product development. This is often referred to as pure research for the development of new sciences and methods.

Breakthrough projects are those projects that will involve a significant change in the product and or process in order to establish a new core product or process family that may result in several new products being introduced into the marketplace. These projects typically result in significant advances in market share or changes in market position for the business.

Platform projects focus on a known product family within the organization that will be utilized as a base for the design of new products or processes that are directly related to that platform. An example of this is shown in a typical automaker, where several different makes and models of cars are based upon the same platform. This simplifies the basis of development of these complex systems.

Derivative projects are typically focused on cost reduction or product or platform enhancements for the development of more streamlined and cost-effective production processes. These projects are usually undertaken in incremental improvement efforts within the organization, and they are intended to keep the existing product line viable and profit generating for the near-term.

Many resources exist that delineate the responsibilities of the Program Manager. One of the most complete of these can be found at the Defense Acquisition University website (www.dau.mil). This document, while being focused on acquisition programs for the government, nonetheless contains extremely valuable information for the Program Manager in any commercial organization.

The first area of responsibility for Program Management is the acquisition process and resources available. This means the Project Manager must understand the mechanism by which funding is allocated, how resources are applied, contracting methods, and the development lifecycle of a project or program. For example, the Program Manager must be familiar with the five phased approach to long-term program developments, Including; concept development phase, technology development phase, product development phase, deployment and sustainment phase, and phase-out. Without a clear understanding of how the Program Management process is intended to function, a typical Program Manager will find it challenging to accomplish the program's objectives.

Lifecycle logistics is also a crucial subject area, focusing on managing the program from cradle-to-grave, or through the entire lifecycle of the program. This includes a great deal of planning, including the development of lifecycle sustainment metrics, affordable system operational effectiveness, a product support strategy, integrated product development, and a model describing how product logistics will be accomplished within the business.

Sustainment issues are also typically managed within the program. These efforts are focused on planning for spiral developments and enhancements to the system based upon the typical development cycle of the technologies employed. For example, when one is producing a ship for the U.S. Navy that will typically last thirty to thirty-five years, it is clear that computing systems located within the ship have a typical lifecycle of eighteen months and will need continuous refresh efforts associated with their constituent elements.

Program Management is also focused upon the organization of the program office, with varying structures being employed. These will include a traditional management structure, functional structure, pure product structure or a matrix orientation. In general, the most widely excepted construct in today's development environment is the use of integrated product teams or IPTs, which are a variation of a matrix organization.

The role of manufacturing in the acquisition process is typically defined very early, and is associated with defining the appropriate production methods and processes that will result in greatest benefits of the program. This will also typically cover the domain of producibility, which will consider the role of quality management systems and design for manufacturability in the early phases of the program development.

Testing evaluation is defined throughout the program. This will include factors associated with modeling and simulation planning, and how those models and simulations will establish the performance of the system in comparison with the expectations or needs of the stakeholders.

Program Managers are also fundamentally responsible for the appropriate planning and scheduling and control of the programs and projects under their umbrella. This will include many of the planning tools previously described, whereas Program Managers are often required to educate their various project teams in order to establish a standardized process and approach to planning and execution of projects. The standardized approach will enable the Program Manager to establish an appropriate business rhythm and reporting structure that will simplify the oversight requirements of the multitude of projects feeding into the program.

Risk Management is a key area for Program Managers, as they must establish the appropriate risk undertaken by the program such that they can maximize the return on investment for the organization. Included in this area are the definition of technical performance measurement metrics within the program, a Risk Management process, and various tools such as Root Cause Analysis, Fault Tree Analysis and Failure Mode Effects and Analysis. These can be used extensively to establish direction for the continued development of technologies on the program. The Risk Management process is a continuous process that will become the principal focus of the Program Manager during the execution of the program.

Systems Engineering is a large part of the Program Manager's domain responsibilities. This includes the systematic development of the users and user needs and system requirements prior to the design of the system. A single chapter in this text is dedicated to the domain of Systems Engineering, and these concepts will be discussed there.

Program Managers must also possess excellent leadership and managerial skills. These are necessary to properly organize and manage the program management office as well as to delegate work amongst the resources provided. Program Managers must also be adept at coaching their direct and indirect reports and managing effectively within the integrated product team environment. By virtue of these needs, the Program Manager must clearly and concisely communicate the expectations, and how performance will be measured with regard to the program. Normally this is accomplished by means of a detailed team charter as well as generalized skills such as working effectively with groups, getting high performance from development teams, running effective meetings, employing effective communications, effectively managing time and utilizing continuous process improvement techniques.

Implications for the Leader

It is vitally important for the Engineering leader to understand the roles and responsibilities of the various members of the program team. The Engineering leader must ensure that the Program Manager is accomplishing those tasks falling within the Program Management responsibilities, such as requiring that appropriate vision and strategy be provided to the project teams.

Secondarily, the engineering leaders must be sure that the projects executed within the engineering domain are supportive of the program management organization and expectations. Therefore, although the engineering leader will typically require status updates of his or her project teams, they must be designed to allow them to be utilized by both the engineering department and the Program Manager. Clarifying these expectations and methods effectively for the team can greatly simplify the efforts and burden placed on the team by management and program. A description of appropriate roles and responsibilities for different stakeholders on a program or project is included for reference.

Roles and Responsibilities

The roles and responsibilities are defined for each of the typical roles on a project. These roles are then defined with regard to the typical tasks and responsibilities to be accomplished during each of the typical life-cycle phases in a summary view. The specific role of the sponsor is clarified in further detail throughout the remainder of the document. These typical project roles include:

- Organizational Management
- Project Sponsor / Business Sponsor
- Program Manager
- Project Manager
- Project Team
- Customer

(from USDA Project Management Handbook, Version 1.0, online publication)

Organizational Management

Organizational Management is responsible for the identification of the need and opportunity for a project, assessment of project risk, and the approval of the project's feasibility and resources. It must also establish the strategic plans and validate that projects are consistent with customer and organizational requirements. Management provides close oversight for high risk or high cost projects.

MANAGEMENT ROLES AND RESPONSIBILITIES					
General	**Initiating**	**Planning**	**Executing**	**Monitoring/Controlling**	**Closing**
Provide leadership and resources to establish and improve project management Ensure that sufficient resources are available to conduct projects Review and approve commitments to external entities (e.g., customers, vendors) Ensure staff is properly trained in project management techniques and principles	Select Project Manager and assist in Project Team staffing Review, validate and approve project charter Authorize and provide funding	Verify that project goals and objectives are defined Review and approve Project Management Plan, cost, risk and establish management reserves Provide management oversight as predicated by review of the project risk analysis, risk response planning and Project Management Plan Enable project staff availability	Regularly conduct executive management reviews and provide oversight	Review project status and corrective action plans (if required) Review and approve changes affecting scope, timing, cost, and/or quality, as required	Validate project completion (goals & objectives) Verify customer and sponsor acceptance Review and close project accounting/ financial files Review project lessons learned and post project reports for continuous improvement action

Project Sponsor / Business Sponsor

The Project Sponsor is usually a member of the management team who will be the recipient of the project's end result (the product). The Project Sponsor is typically the head of a program area. This individual makes the business argument for the project to exist, controls the overall funding of the project and defines the acceptance criteria of the product.

SPONSOR ROLES AND RESPONSIBILITIES					
General	**Initiating**	**Planning**	**Executing**	**Monitoring/Controlling**	**Closing**
Articulate project and/or customer requirements Validate that project requirements are met Provide the necessary funding and resources as appropriate Champion the project to provide exposure and buy-in Communicate the sponsor's views on project progress and success factors to the Project Team and other stakeholders Define project goals, objectives and success criteria	Provide the strategic goals and objectives of the recipient organization and guidance to the Project Team to identify the relevance and value of the project Obtain or provide funding for the project	Review and approve the Project Management Plan and management approach Participate in planning sessions	Attend executive requirement reviews Resolve escalated project requirements issues, remove barriers and obstacles to the project Provide written agreement to project requirements and qualifying criteria	Attend and participate as needed at Project Status Reviews and steering meetings Attend change control meetings and review and approve changes in scope, timing, quality and/or cost as impacted	Provide representation or input to lessons learned reviews Sign off on project completion

Program Manager

The terms "program" and "project" management are often used interchangeably. However, within this document the two terms and concepts are separate and distinct. Program management is defined as a group or series of related projects and ongoing systems/applications managed in a coordinated way to achieve resource, cost and quality efficiencies not available to individual projects. Programs generally support strategic goals and objectives, while projects may be more targeted in focus. The Program Manager has responsibility for the management of a series of related projects and the management of the corresponding Project Managers.

PROGRAM MANAGER ROLES AND RESPONSIBILITIES					
General	**Initiating**	**Planning**	**Executing**	**Monitoring/Controlling**	**Closing**
Plan, organize, staff, direct, control and coordinate Recommend composition of own program team Own and guide the program Reward and recognize performance Be accountable for cost, schedule, quality and scope Resolve any outstanding issues among the Project Teams that cannot be resolved within the team Responsibility for overall resource allocation for Project Managers assigned to the program Maintain ongoing	Assign Project Manager and assist in Project Team staffing Review/validate/concur on project charter Validate and communicate individual project objectives	Verify that project goals and objectives are defined Verify that project is aligned with the strategic goals of the program Review/concur with Project Management Plan, cost, risk and establish management reserves Provide guidance on cost and schedule development Ensure project staff availability Conduct routine program planning sessions as	Conduct regularly scheduled project reviews	Review project status and corrective action plans (if required) Review, concur and participate in milestone review briefings Review/concur on changes affecting scope, timing, cost, and/or quality, as required Prioritize any changes to project scope	Review whether stated improvements or benefits were realized from the project Assure customer and sponsor acceptance is obtained Review and concur on project accounting/ financial file closeout documents Review project lessons learned and post project reports for continuous improvement action

communication with the Project Managers from the program management level perspective Communicate project status to fellow Program Managers Ensure applications are developed consistently with the project management cycle Ensure projects are managed in accordance with the recommendations for project management as outlined in the Project Management Handbook Ensure that security certification and accreditation (C&A) requirements are met.		defined by individual's organization			

Project Manager

The Project Manager has overall project responsibility. In order to achieve success, the Project Manager should work closely with the Sponsor with respect to staffing requirements and funding availability. The Project Manager is responsible for completing the project on time, within budget, and meeting the quality criteria and requirements.

The Project Manager should be assigned as early as possible in the life cycle of the project in order to establish project ownership and management responsibility as well as to begin the development of the project requirements from the "ground up."

PROJECT MANAGER ROLES AND RESPONSIBILITIES					
General	**Initiating**	**Planning**	**Executing**	**Monitoring/Controllin g**	**Closing**
Comprehend and implement organizational project policies and procedures Maintain project staff technical proficiency and productivity, and provide training where required Establish and maintain project quality Identify, purchase and acquire project infrastructure needs Identify and secure Project Team resources Serve as focal point for project communication s	Develop project concept document Define sponsor and organization s needs Develop Project Charter and obtain approval Identify and document project constraints Identify and document project assumptions Document requirements Review customer requirements for	Develop Project Plan, including Project Charter, Scope Statement, constraints, assumptions, WBS defining project deliverables, cost estimates and project budget, major milestones, schedule, resource requirements, acquisition/procuremen t plans, risk analysis and response plans, Project Team structure and communications plan. Also included will be the deliverables acceptance criteria (quality metrics) and the acceptance process. Develop the supporting plans such as scope, cost, risk, schedule, quality, resource, security deliverables, procurement and change management plans Obtain stakeholder	Manage and monitor day-to-day activity and provide direction to team members and supporting organizations Manage to and monitor quality targets and goals (both project and product) Manage and monitor risk response strategies Disseminate project information and maintain communicatio n Develop and update system security plan and other	Develop and distribute project performance reports. Regularly review project status, evaluating performance criteria (scope, cost, schedule & quality) Develop and manage corrective action plans Evaluate project performance and initiate change requests as required (scope, cost, schedule or quality) Participate in change control board to review and approve product/project changes Review project risks and establish risk response plans Adjust Project Management Planning, as required, to include approved changes in scope, timing, cost or quality after obtaining customer approval	Obtain customer and management approval and acceptance of completed product Complete contract closure process Close out open action items Develop post-implementation report Conduct lessons learned session and develop recommendation s for continuous improvement Close out any financial accounts or charge codes Archive all project data Recognize

Ensure that security requirements are met	feasibility and available resources	approval and acceptance of the Project Management Plan Obtain organizational commitment and support for completion of project task assignments, timing and quality Establish baselines	security deliverables Manage, or support, purchases and acquisitions process and contract administration requirements		Project Team and celebrate success

Project Team

The Project Team is responsible for performing the project activities. Project Team members, as required, may assist the Project Manager with planning the project (e.g., scope and WBS) and they may also assist with obtaining commitments to complete the project within established schedule and budget constraints. Customers and/or Stakeholders should interact with the Project Team to ensure that requirements are properly understood and implemented.

PROJECT TEAM ROLES AND RESPONSIBILITIES					
General	**Initiating**	**Planning**	**Executing**	**Monitoring/Controlling**	**Closing**
Identify product alternatives Complete the project within budgeted cost, schedule and quality requirements Support Project Management Planning and control Participate in identifying, mitigating, and monitoring project risks	Provide estimates for product deliverables Analyze requirements for clarity, completeness and consistency	Develop technical approach Participate in the development of the Project Management Plan Participate in Scope Planning Identify tools needed for project Identify staff training needs	Create product and process solutions Conduct or participate in internal and external reviews Provide quality assurance support Manage work effort to maintain on time, on cost and on quality results Identify any project roadblocks, barriers or unanticipated risk events	Track the project execution effort and submit status reports (scope, cost, and schedule) Maintain project and product quality requirements Identify and react to risk events as they are identified or occur Participate in change control reviews	Participate in lessons learned sessions Identify ways to improve project processes or products (continuous improvement) Turn over all project related documentation to the Project Manager for archiving

Customer

A Customer is responsible for communicating project needs and verifying that requirements have been met at project completion. A Customer may be internal, external or both.

CUSTOMER ROLES AND RESPONSIBILITIES					
General	**Initiating**	**Planning**	**Executing**	**Monitoring/Controlling**	**Closing**
Articulate customer requirements Validate that project requirements are met Support and conduct staff training programs as required to make certain that the staff is "ready to accept" the new product Be proponent of the new product to the customer organization	Clearly define customer needs and requirements to the Project Manager and Project Team	Review and approve Project Management Plan Attend and participate in project requirement reviews Assign customer personnel as project points of contact Provide written agreement on requirements and qualifying criteria Provide input into deployment planning	Review project status reports Provide project support infrastructure as required	Participate with Project Team developing corrective action plans addressing variances in time, cost or quality Communicate identified issues associated with project or product performance Validate quality assurance of deliverables Participate in change control process Review and approve or escalate project changes affecting scope, time, cost or quality	Accept the completed product Provide representation or input to lessons learned reviews

Chapter 5: Systems Engineering Culture

Systems Engineering is a relatively recently developed domain of knowledge that is intended to improve the typical engineering organization's ability to deliver products that will meet customer needs more readily than those which have been developed in the past.

Systems Engineering does not require the development of skills in a new science or mathematical field. This makes it unique in engineering domains. By virtue of this, many different disciplines can develop capabilities that will result in excellent Systems Engineering approaches to problem-solving, although in general the majority of Systems Engineers are also Mechanical or Electrical or Software (or other) Engineers as well.

Simply stated, the system is an integrated composite of products, components, people and processes that will provide capabilities that will satisfy the needs of a customer.

Systems Engineering consists of two significant areas of interest: the technical knowledge domain in which the Systems Engineer operates, and Systems Engineering Management. Therefore, Systems Engineering is an interdisciplinary engineering management process that develops and verifies an integrated set of system solutions that will ultimately satisfy customer needs, whether implicit or explicit.

It is imperative for the engineering leader to utilize an effective subset of Systems Engineering tools and techniques in order to ensure efficiency and provide focus in the development of solutions for customer needs. This section of the text is not intended to be an exhaustive description of all elements of the Systems Engineering domain of knowledge, but rather a quick overview to provide some guidance for the interim leader to begin to incorporate Systems Engineering concepts into his or her product development process.

The Systems Engineering Approach

Many different models have been put forward with regard to the development of a system utilizing Systems Engineering principles. One of the most common has been adopted by INCOSE (the professional organization for Systems Engineers), which has been adapted from many authors in the domain including the Defense Acquisition University. In general, these process documents are quite complex and challenging to understand, but when boiled down to simple processes and a step-by-step methodology for product development, they are easily understandable and applicable for virtually any product development organization. The V diagram of Systems Engineering is provided in Figure 39.

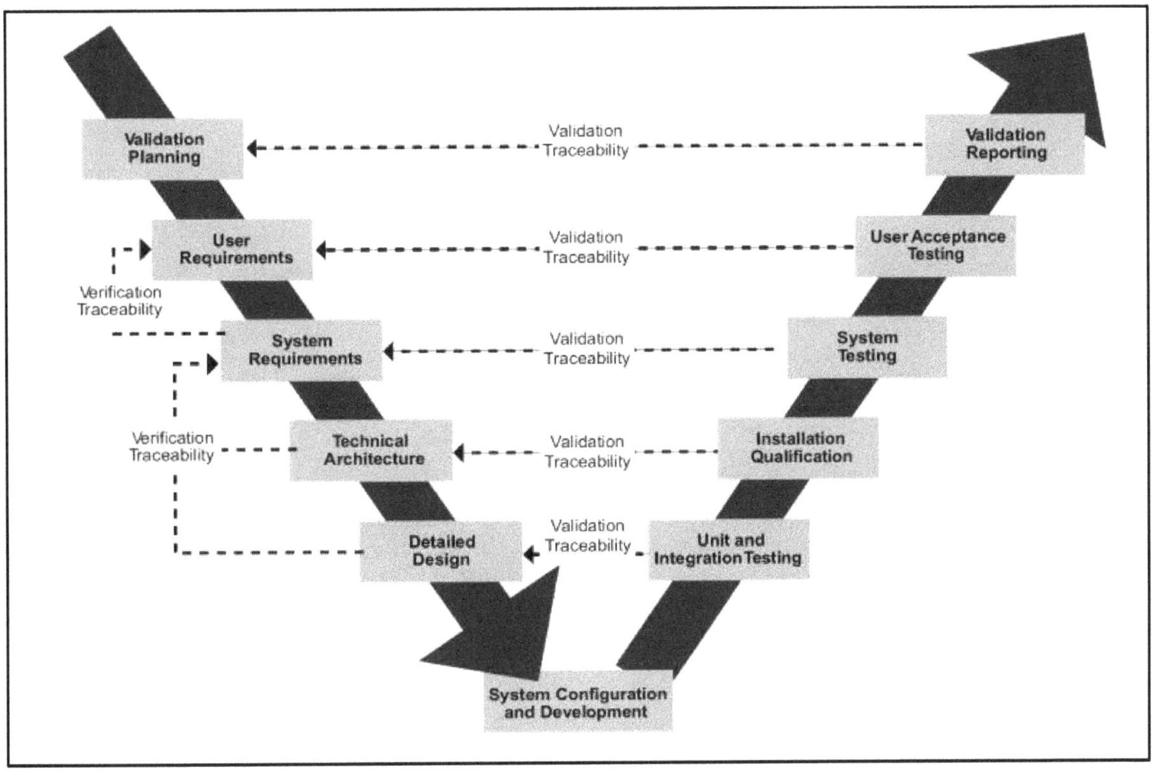

Figure 39: Systems V Diagram

This diagram demonstrates from a simplified perspective how the systems process will typically be applied in an engineering development project. Beginning from the upper left, the Systems Engineer will define users, user needs and ultimately (utilizing the tools in Systems Engineering methodologies), the requirements necessary for the project. Once the requirements have been developed, the Systems Engineer will continue to develop a systems architecture, which will ultimately be translated into a physical architecture and a detailed design, much as any standard engineering process is conducted today.

This simplified version was derived from a more complex version provided by the

Defense Acquisition University. In this version, the entire program management cycle is included, and spans from the initiation of a program through the final retirement of the program products. This construction can be seen in Figure 40.

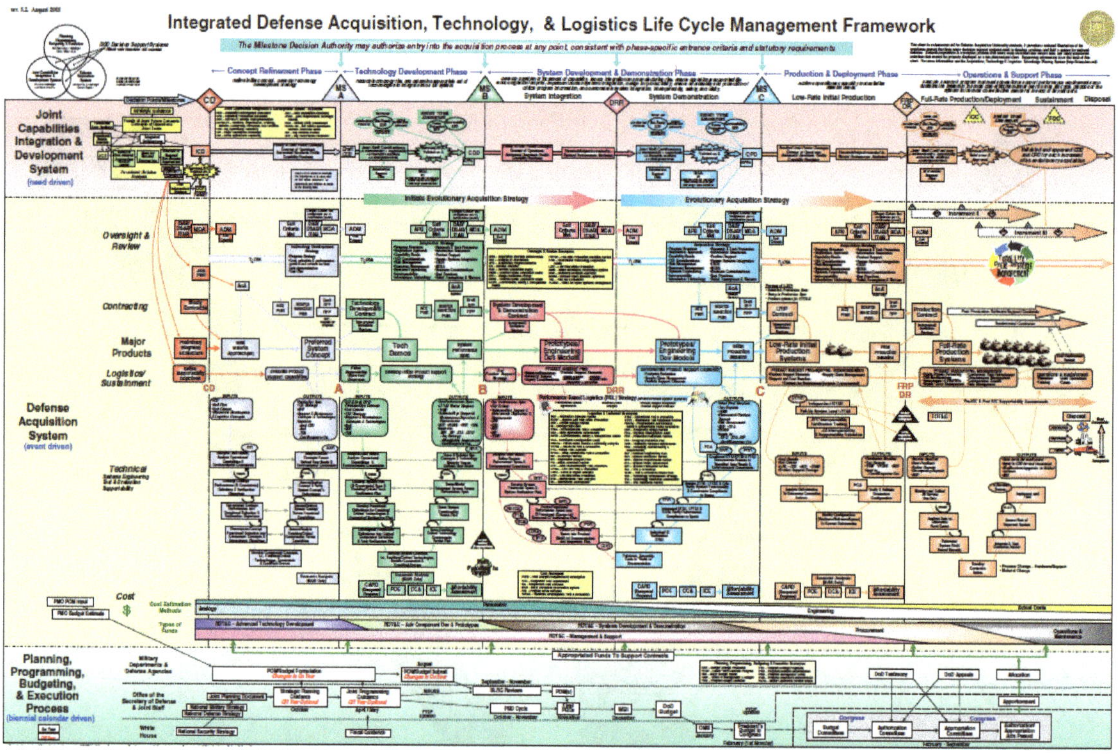

Figure 40: DAU Systems Process (www.dau.mil)

This is a highly complex diagram, and is often well beyond the needs of a typical engineering organization. However, upon reflection and observation one can readily understand that this diagram is nothing more than the combination of five independent System Development cycles in a longitudinal format. Therefore, we can investigate a much smaller section of this diagram and note the typical the diagram of systems development. This is shown in Figure 41.

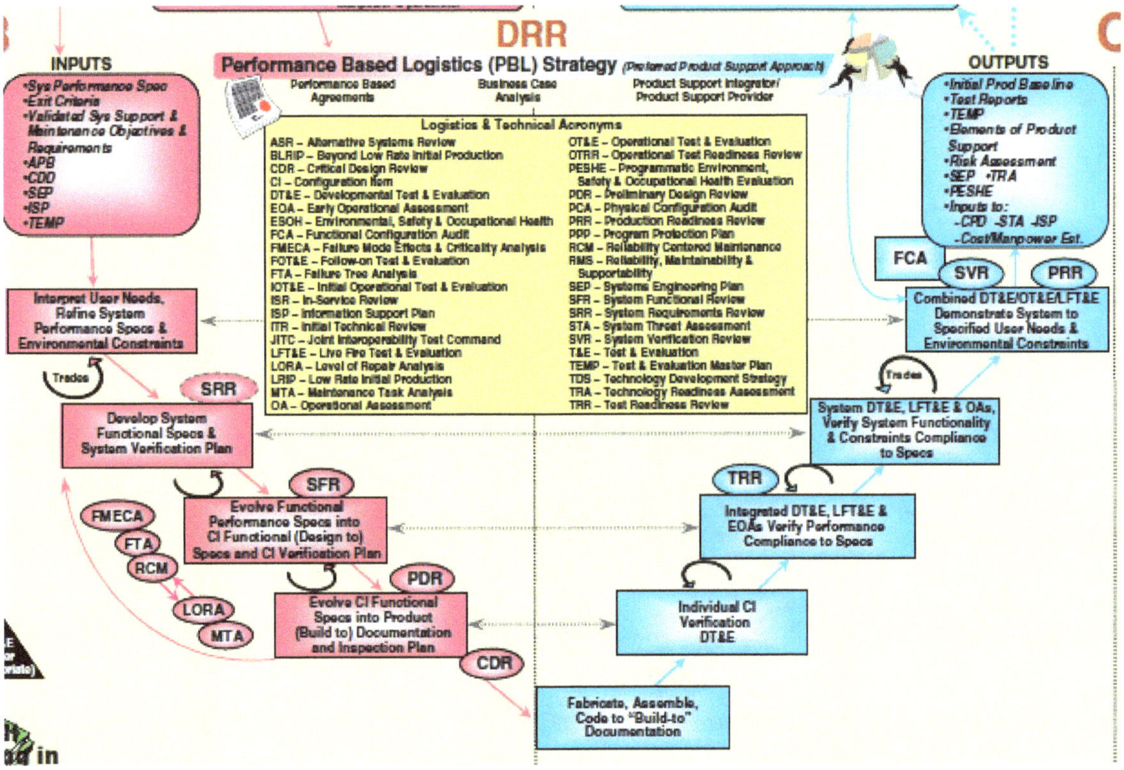

Figure 41: DAU Product Development Process

As can be seen, this process closely resembles the previously described model of development, so engineering leaders are urged to investigate this highly detailed model of the Systems Engineering process and choose those elements of greatest impact for their engineering organization. A subset of this complex system has been selected for discussion in this text, and has proven to be an effective methodology for small to midsize manufacturing and product development companies.

Systems Engineering in Context

To understand how Systems Engineering differs from Mechanical or Electrical Engineering, we will first examine a typical project schedule. This particular work breakdown structure has been developed as a generic WBS for a typical manufacturing entity. The reader will note that the system described contains mechanical, electrical, and software components.

A common approach to developing a new product with this content is shown in Figure 42.

Task Name

- NPD Project Template
  - Gate#1 Project Approval
  - Develop Concept
  - Gate#2 Approval to Engineer Product
  - Product Design Phase
    - Mechanical Design
      - Develop Layout
      - Define Major Components
      - Develop Detail Design
      - Develop Assembly Drawings
      - Develop BOM
      - Check Drawings
      - Correct DWGs as needed
      - Release for Initial Pilot
    - Electrical Design
      - Develop Layout
      - Define Major Components
      - Develop Detail Design
      - Develop Assembly Drawings
      - Develop BOM
      - Check Drawings
      - Correct DWGs as needed
      - Release for Initial Pilot
    - Software development
      - Layout System
      - Develop ICDs
      - Define Major Components
      - Code Components
      - Initial testing
      - System Refinement
      - Preliminary release for Pilot
  - Gate#3 Approval to Build Prototype
  - Build Prototype
    - Execute In-House Testing (startup and verification)
    - Revise Design as Required
  - Gate#4 Approval to Release for Production
  - Production Release
  - Production support

Figure 42: New Product Development WBS

In order to incorporate a Systems Engineering expertise into this WBS, we must add additional Systems Engineering tasks in two specific areas. These include the incorporation of systems development tasks prior to the detail engineering design of the system, as well as verification and validation tasks that will be incorporated in place of testing. These tasks will be inserted in our project plan as shown in Figure 43.

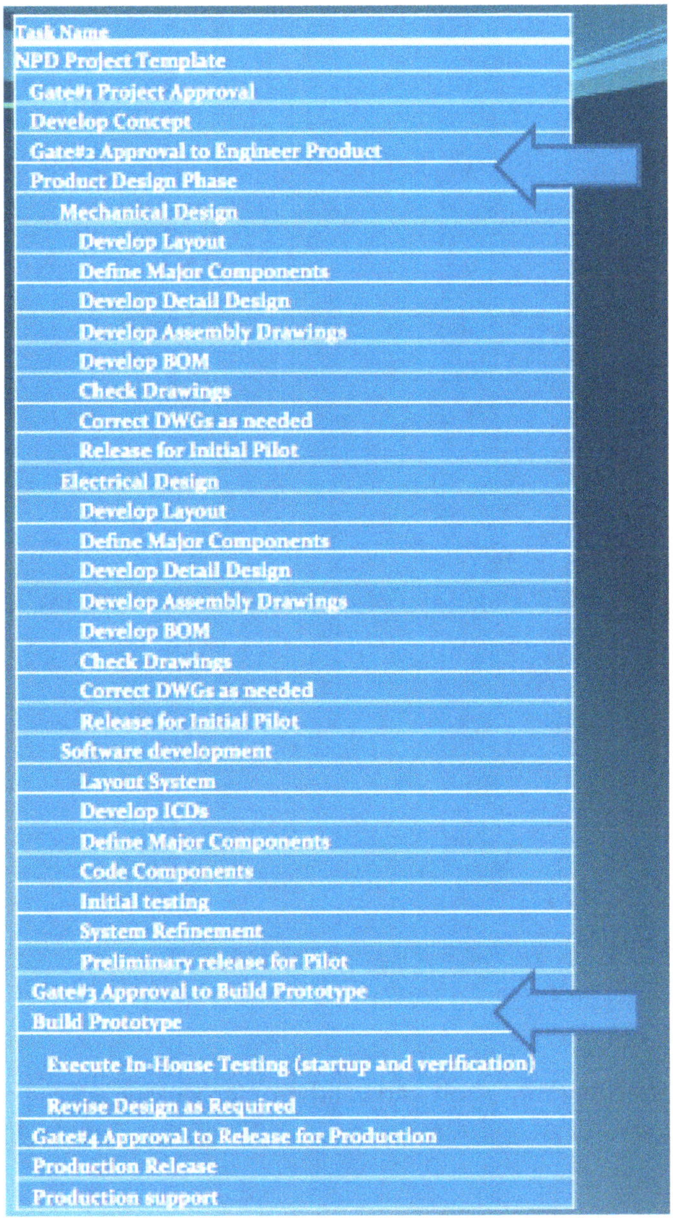

Figure 43: Where Systems Engineering augments a Project

The tasks that will be incorporated representing the Systems Engineering development are as follows:

- Product Design Phase
  - Detailed Project Planning
    - Develop Specification
    - Develop Schedule
    - Develop Budget/Risk and Charter
  - Systems Engineering
    - Identify Users
    - Identify User needs
    - Define development team
    - Develop AV-1
    - Establish general system requirements
    - Develop OV-2
    - Develop OV-5
    - Refine to black box level
    - Establish RVTM document
    - Populate requirements from views
    - Define System scope
    - Develop functional analysis
    - Compare OV-5 to functional analysis
    - Establish System Architecture
    - Define roles and responsibilities for physical architecture
    - Conduct Preliminary Design Review
  - Develop detail design
    - Mechanical Design

Figure 44: Systems Development Tasks

Those additional tasks incorporated for the verification and validation process are as follows:

Verification and Validation
Develop V&V plan
Execute In-House Testing (startup and verification)
Startup & Debug
Performance Testing
Revise per Test results
Field Beta Testing (Validation)
Tear Down Machine
Ship Machine
Install Machine
Field Testing of Machine
Revise Design as Required
Mechanical Revisions
Electrical Revisions
Conduct Critical Design review
Gate#4 Approval to Release for Production

Figure 45: Verification and Validation tasks

If a Gantt chart is utilized to demonstrate a generic project schedule with systems tasks, it would look similar to Figure 46. Note that the previously identified tasks have been added to a standard WBS in order to adopt a Systems Engineering perspective.

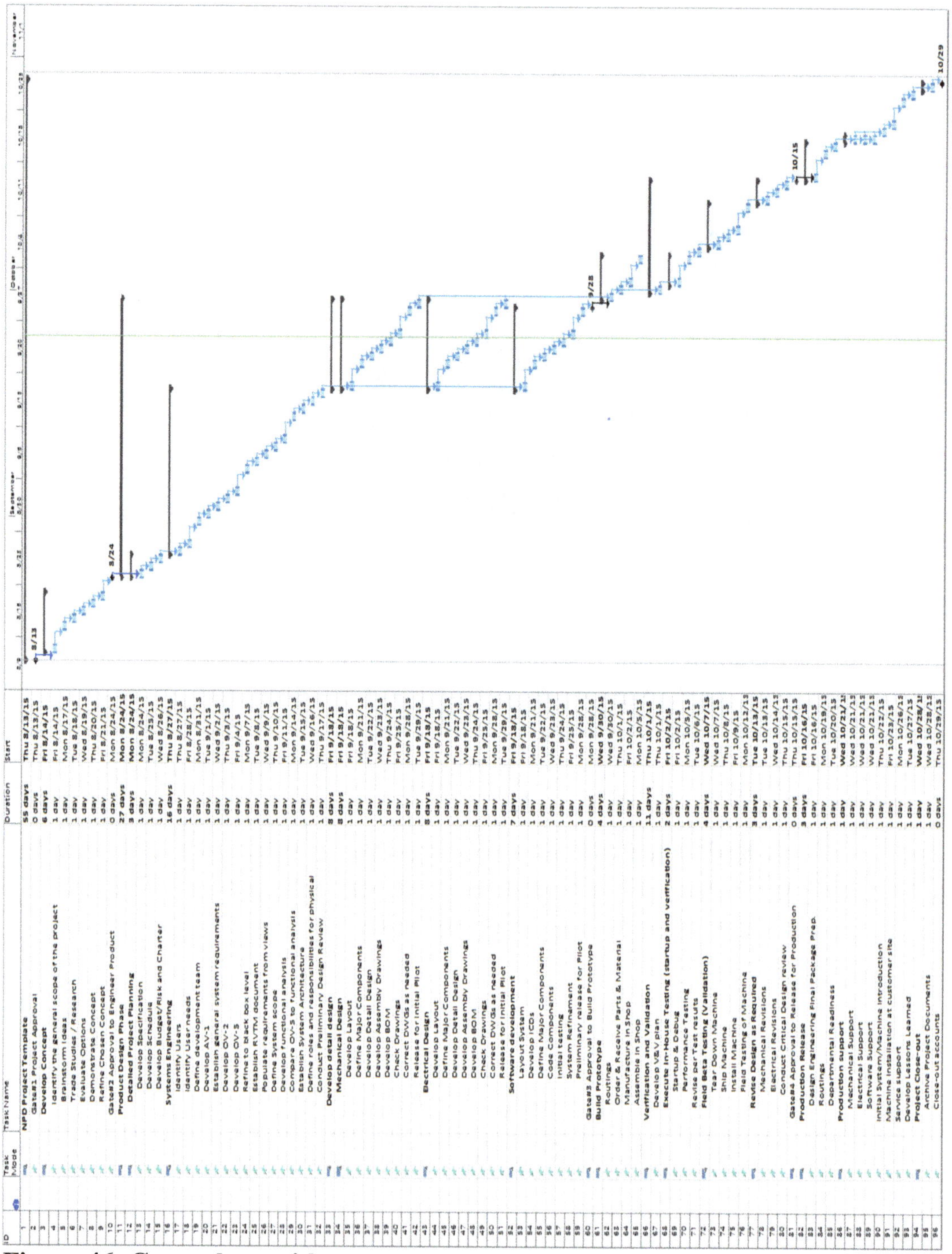

Figure 46: Gantt chart with systems tasks

The Systems Engineering Process

There is a distinct difference between a classically trained Engineer and a Systems Engineer. In general, this distinction revolves around the ability of the Systems Engineer to develop a broader understanding of a problem and context within which the problem exists. Systems development in general requires a deeper understanding and a more reflective approach to formulating the problem than does classical engineering.

Therefore, Systems Engineers must consider more than the typical Engineering approaches to problem solving. In fact, many of the system tools are constructed in a manner that enables us to develop a more complete understanding of the issues of importance to the customer and developer. By more fully understanding the situation and problem, we can better address the problem and be more effective and efficient in the application of our engineering resources.

One considerable challenge that Engineers and Scientists face is the fact that we are typically intelligent and knowledgeable about the problem at hand. This knowledge and intelligence, when coupled with significant experience, often results in a quick jump to what we believe to be the optimal solution to the problem. In short, many engineers will look at a problem and almost immediately envision a solution to that problem. This is problematic for Systems Engineers, as it very quickly limits our ability to expand our understanding of the problem, and keeps us thinking 'inside the box', so to speak. Therefore, much of our development approach will be focused on keeping the Systems Engineer away from detail design as long as possible, until we have fully completed our systems architecture development. If we can accomplish this, we can remove initial bias and limitations from our final design, thereby creating a more focused and coherent solution to the problem not limited by our preconceived notions.

To understand the challenges we face, we will undertake a simple problem to illustrate these issues for our consideration. Imagine that you have been asked to develop a display for our local art museum, to put on display in our permanent collection a new and exciting acquisition. What information is necessary? When does the engineer fully understand the essence of the challenge?

Many Engineers would ask initial questions regarding the new acquisition. For example:
Is it a painting or a sculpture?
How much does it weigh?
What is its value?

How big is it?

If we establish that this is an approximately life-sized sculpture of a human being, carved from marble and weighing approximately six thousand pounds, most engineers have enough information to envision the final product.

If asked to document that final product, the most common solution would include a pedestal base reinforced to support and position the mass of the statue, which will be located on the floor of the museum and painted black. This is the extent of the system's development, and this is the solution generally provided.

Is this truly an adequate solution? Have we fully addressed the needs of the customer, and have we understood the true problem?

As a Systems Engineer, one tool that is utilized to enable us to better understand our problems is to look at our challenge from multiple viewpoints. These viewpoints might include different physical perspectives, different functional perspectives, knowledge domains, or a variety of other approaches. To explore these concepts, we will examine a photo of the statue in question.

Figure 47: Photo of museum acquisition

This new viewpoint immediately provides us with better understanding of the problem. Now we become more aware of the age of the acquisition, and perhaps the delicacy of the system that we must be aware of in our museum display.

Does this fundamentally change the display we have previously designed?

In most cases, we feel that the originally proposed design is still a superior solution, although we may incorporate slight enhancements and modifications to the design.

But what if a new viewpoint is provided? What if we consider the historical provenance of this acquisition? This information is provided as follows:

Location: Louvre Museum, Paris, France

Artist/Maker: Unknown

Description, English: So-called "Venus de Milo" (Aphrodite from Melos). Parian marble, ca. 130-100 BC. Found in Melos in 1820.

Dimensions: H. 2.02 m (6 ft. 7 ½ in.)

Credit: Gift of the Marquis de Rivière to Louis XVIII of France, 1821

Accession number: Ma 399 (LL 299)

Current Location: Temporary - Department of Paintings, Sully, first floor, room 74

References: Hamiaux, M., Les Sculptures grecques, II, Paris, 1998, no. 52, pp. 41-44

This additional viewpoint will often prompt distinct changes in the design previously established. Now the designer will frequently focus on security issues, care for the statue, and potential insurance for liability issues. Therefore, by enhancing the understanding of our challenge with insight into the acquisition, we have dramatically enriched our approach to the problem. This is in essence what the Systems Engineering tools enable the Systems Engineer to accomplish.

Before moving on from this challenge, let us consider many more viewpoints of the Venus de Milo statue, focusing on what insights these might provide us as Systems Engineers, and subsequently what changes might be made in our technical approach to this challenge.

Figure 48: Images of Venus de Milo

From these images, most Engineers would observe that lighting is clearly a significant factor, the background provided to the statue is of considerable importance, the ability of observers to see the statue from multiple angles is also of significant value. The statue is already exhibiting significant damage and decay, and therefore must be protected carefully when placed on display. Additionally, questions of security and safety for the display are often considered.

Again we see that the addition of information provided from different viewpoints can enrich our understanding of the problem significantly and therefore change the technical approach or solution.

If we take one final pass at increasing our understanding of this problem, we could observe the Venus de Milo on display at the Louvre in Paris, France. Upon observation, we would note several things:

The statute is located in a large salon
No other artworks are located in the same salon
The statue is raised a significant height above the floor
The security barrier is approximately the same height as the pedestal
There are a large amount of people that come to view the statue
Many people would like their picture taken with the statue
The statue is more of a destination than an appreciated artwork

Upon reflection of this added data, it becomes clear that the challenge of placing this statue on display goes far beyond the design of a stable platform on the floor of the museum. We realize the real challenge is to create a setting that allows for effective movement of people throughout the museum and around the statute, free access to take photographs, and protection of the statue from the environment and the museum visitors. Due to the large number of sightseers, we must also provide additional cooling through the museum's heating and ventilation system.

If we had approached the problem based upon our initial assumptions and conclusions, it is unlikely we would have produced as an effective a solution as we ultimately arrived at after considering multiple viewpoints. Therefore, we will explore several methods that are typically used in Systems Engineering to develop multiple viewpoints of the problem at hand.

Beginning Systems Development

We will begin our Systems Development in much the same way that we started our Project Development. This will include the definition of several developmental areas related to the system, including:
Users
User Needs

Generalized System Requirements
Derived System Requirements

In order to complete our list of requirements, we must utilize visualization tools and methods to ensure that we cover an appropriate breadth and depth of system understanding. To address this, we move on to the subject of Architectural Frameworks.

DoDAF views

The Department of Defense has traditionally been involved in extremely large-scale and complex development projects. Because of this, they are often at the forefront of new developments in Systems Engineering methodologies. One of the most powerful tools yet developed for use in Systems Engineering is the methodology recommended for the production of an Architectural Framework. This is known as the Department of Defense Architectural Framework, or DoDAF.

This framework is intended to define a common approach for describing, presenting, and comparing enterprise architectures. It facilitates the use of common principles, assumptions and terminology in any description of the system to be designed. The principal objective of developing a framework of this sort is to ensure that architectural descriptions can be compared and related across organizational boundaries, national boundaries, and even within organizations.

One critical factor in applying this framework methodology is to focus upon the system to be developed as an element of an enterprise. This means it is no longer acceptable to design a machine or a device that is completely stand-alone and independent of all other systems with which it will come into contact. If we treat the system as an enterprise, we are no longer simply designing a new aircraft; we are in fact designing a new vehicle for use in the air system. This mindset is very effective for maintaining a coherence and pragmatic development strategy in most organizations.

An Architecture is defined as the structure of components, their relationships, and the principles and guidelines governing their design and evolution over time. An Architecture is the fundamental organization of a system embodied in its components, their relationships to each other and within the environment, and the principles guiding its design and evolution.

The intent of System Architecture is to allow appropriate make/buy decisions to be made prior to the development of systems or components, the ability to discriminate between options on a development project, the development of true and complete requirements, and to drive one or more systems to a common use or purpose.

System Design, on the other hand, is used to develop the system components, build the system, and manage the changes and modifications to the system during the development cycle.

The enterprise's competitive edge and ultimate success are enabled by its ability to rapidly respond to changing business strategies and technologies. The competitive edge translates into higher levels of customer satisfaction, shorter work cycles, reduction in schedules, reductions in maintenance costs, and reduced development time, which all result in lower overall lifecycle costs.

Enterprise Architecture is the key facilitating ingredient providing a holistic view and a mechanism for enabling the design and development of innovative systems, as well as providing for a common communication platform and method to enhance the understanding of the enterprise. The overarching goal of Enterprise Architecture is to manage the complexity of the enterprise, online business strategies and implementations, and facilitate innovations and rapid change in order to maintain business and technical advantages in the marketplace.

Any Architectural Framework is basically a tool. It should describe the methods for designing a system in terms of a set of building blocks, and for showing how the building blocks fit together. It should contain a set of tools and provide a common vocabulary.

The DoDAF work products are principally organized around three independent views, and also include a fourth summary view that is used to consolidate the other views. These views are described in Figure 49.

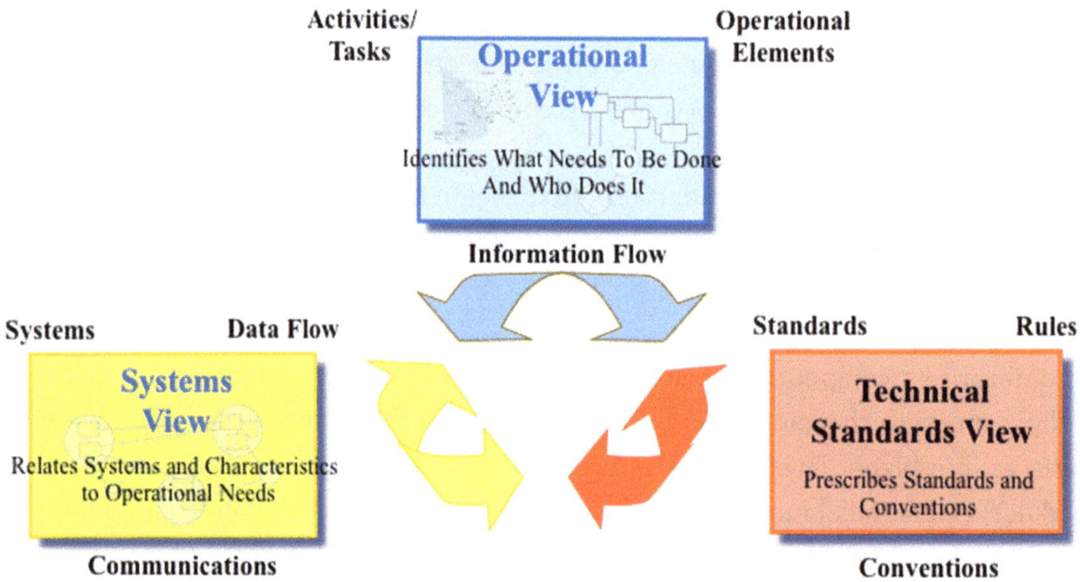

Figure 49: DoDAF work products

The DoDAF describes a set of twenty-six different work products that are used to ensure uniformity and standardization in the documentation and communication of the Architecture. The use of all twenty-six of these work products (or views) enables the documentation of an entire Architecture from requirements through implementation. In general, a smaller subset of work products is typically selected based upon the specific needs of the system being developed.

Therefore, the list of products is defined and categorized into four views:

All Views (AV): the summary information describing the Architecture's plans, scope, and definitions

Operational View (OV): the behaviors and functions describing the way the system is used in operation or by the customer

Systems View (SV): the system and applications supporting the mission functions

Technical Standards View (TV): the policies, standards, and constraints of the system

The entirety of the Framework Work Products are listed in Figure 50.

Applicable View	Framework Product	Framework Product Name	General Description
All Views	AV-1	Overview and Summary Information	Scope, purpose, intended users, environment depicted, analytical findings
All Views	AV-2	Integrated Dictionary	Architecture data repository with definitions of all terms used in all products
Operational	OV-1	High-Level Operational Concept Graphic	High-level graphical/textual description of operational concept
Operational	OV-2	Operational Node Connectivity Description	Operational nodes, connectivity, and information exchange needlines between nodes
Operational	OV-3	Operational Information Exchange Matrix	Information exchanged between nodes and the relevant attributes of that exchange
Operational	OV-4	Organizational Relationships Chart	Organizational, role, or other relationships among organizations
Operational	OV-5	Operational Activity Model	Capabilities, operational activities, relationships among activities, inputs, and outputs; overlays can show cost, performing nodes, or other pertinent information
Operational	OV-6a	Operational Rules Model	One of three products used to describe operational activity—identifies business rules that constrain operation
Operational	OV-6b	Operational State Transition Description	One of three products used to describe operational activity—identifies business process responses to events
Operational	OV-6c	Operational Event-Trace Description	One of three products used to describe operational activity—traces actions in a scenario or sequence of events
Operational	OV-7	Logical Data Model	Documentation of the system data requirements and structural business process rules of the Operational View
Systems	SV-1	Systems Interface Description	Identification of systems nodes, systems, and system items and their interconnections, within and between nodes
Systems	SV-2	Systems Communications Description	Systems nodes, systems, and system items, and their related communications lay-downs
Systems	SV-3	Systems-Systems Matrix	Relationships among systems in a given architecture; can be designed to show relationships of interest, e.g., system-type interfaces, planned vs. existing interfaces, etc.
Systems	SV-4	Systems Functionality Description	Functions performed by systems and the system data flows among system functions
Systems	SV-5	Operational Activity to Systems Function Traceability Matrix	Mapping of systems back to capabilities or of system functions back to operational activities
Systems	SV-6	Systems Data Exchange Matrix	Provides details of system data elements being exchanged between systems and the attributes of that exchange
Systems	SV-7	Systems Performance Parameters Matrix	Performance characteristics of Systems View elements for the appropriate time frame(s)
Systems	SV-8	Systems Evolution Description	Planned incremental steps toward migrating a suite of systems to a more efficient suite, or toward evolving a current system to a future implementation
Systems	SV-9	Systems Technology Forecast	Emerging technologies and software/hardware products that are expected to be available in a given set of time frames and that will affect future development of the architecture
Systems	SV-10a	Systems Rules Model	One of three products used to describe system functionality—identifies constraints that are imposed on systems functionality due to some aspect of systems design or implementation
Systems	SV-10b	Systems State Transition Description	One of three products used to describe system functionality—identifies responses of a system to events
Systems	SV-10c	Systems Event-Trace Description	One of three products used to describe system functionality—identifies system-specific refinements of critical sequences of events described in the Operational View
Systems	SV-11	Physical Schema	Physical implementation of the Logical Data Model entities, e.g., message formats, file structures, physical schema
Technical	TV-1	Technical Standards Profile	Listing of standards that apply to Systems View elements in a given architecture
Technical	TV-2	Technical Standards Forecast	Description of emerging standards and potential impact on current Systems View elements, within a set of time frames

Figure 50: DoDAF Work Products

For most organizations, a reduced list of Architectural Framework Products is appropriate for product development. This includes:

AV– 1: Overview and summary information
AV– 2: Integrated dictionary
OV – 2: Operational node conductivity description
OV – 3: Operational information exchange matrix
OV - 5: Operational activity model

Please see http://aitc.aitcnet.org/dodaf for full descriptions of these products.

AV-1

The AV – 1 is a work product intended to describe the general overview and summary for the system to be developed. A typical AV-1 would include the following information:
Name – name of the system being developed
Architect – names of the developers of the system
Organizations involved– listing of major organizational stakeholders
When developed – date for outset of program
Purpose – Reason for the organization's pursuit of this System Development
Scope – What is included and excluded from the System Development
Mission – Customer's reason for developing the system
Geographical considerations – Location where the system will be implemented
Rules, criteria, and conventions – Context for the System Development Findings; used as a parking lot for issues of note throughout System Development
Tools and file formats – Baseline tool requirements for System Development
Constraints, including any and all constraints required for successful completion of the system
Assumptions – Assumptions accepted by the development team

An example AV–1 for a simple power monitoring system is included in Figure 51.

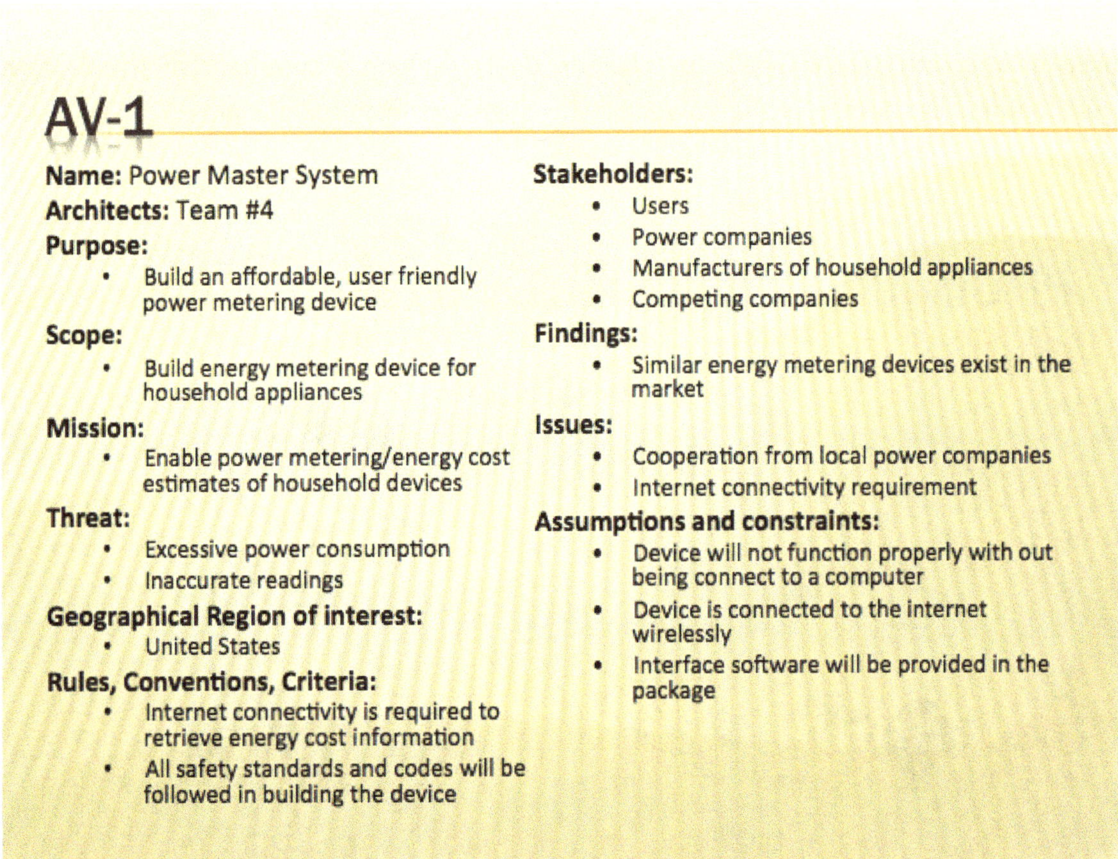

Figure 51: Power Master System AV-1

AV-2

The AV-2 is an integrated dictionary. At a minimum, the integrated dictionary is a glossary with definitions of terms used in a given Architecture's description. Each labeled graphical item in the graphical representations should have a corresponding entry in the integrated dictionary. This provides a baseline of understanding for all members of the System Development team regarding the specific terminology and references used throughout the System Development.

OV-1

The OV-1 is an operational summary of the system. Usually provided in a graphical format, this is essentially a cartoon of the intended system to be developed. One can think of this work product as an abstract for a paper. It is typically developed late in the development cycle, but can be used as a common discussion factor for all members of the System Development team. Therefore, it is useful as a touchstone or guiding philosophy for many members of the development team. Secondarily, it will typically be used as the introductory view for all individuals seeking a summary of the System Development. A sample for the Power Master System is included in Figure 52.

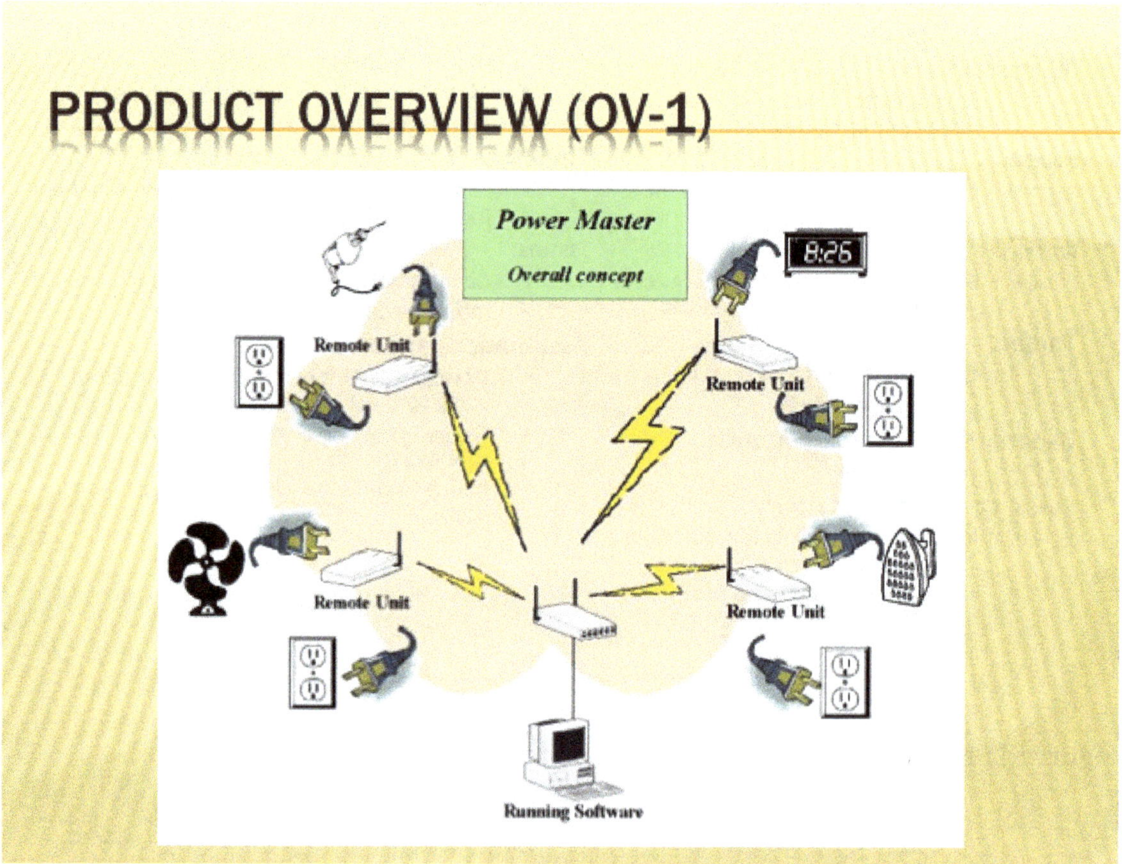

Figure 52: Power Master system OV-1

OV-2

The OV-2 is the operational node connectivity description. It describes how the system interfaces with all other systems and components with which it comes into contact. This may be limited to your enterprise or may extend well beyond it. In general, the OV-2 is constructed by placing the developing system at the center, and connecting all other inputs and outputs to related systems or components. Each input and output is described by a separate arrow and is numbered as a 'need line' for the diagram. These need lines are described on the following page; the Data Elements Table for the corresponding architectural view.

Note that great care is taken in establishing each of the need lines. This is due to the fact that after the architectural review is completed, each of the defined need lines will define for the System Development team one or more requirements necessary for the completion of the System Development.

The Power Master System OV-2 is included in Figure 53.

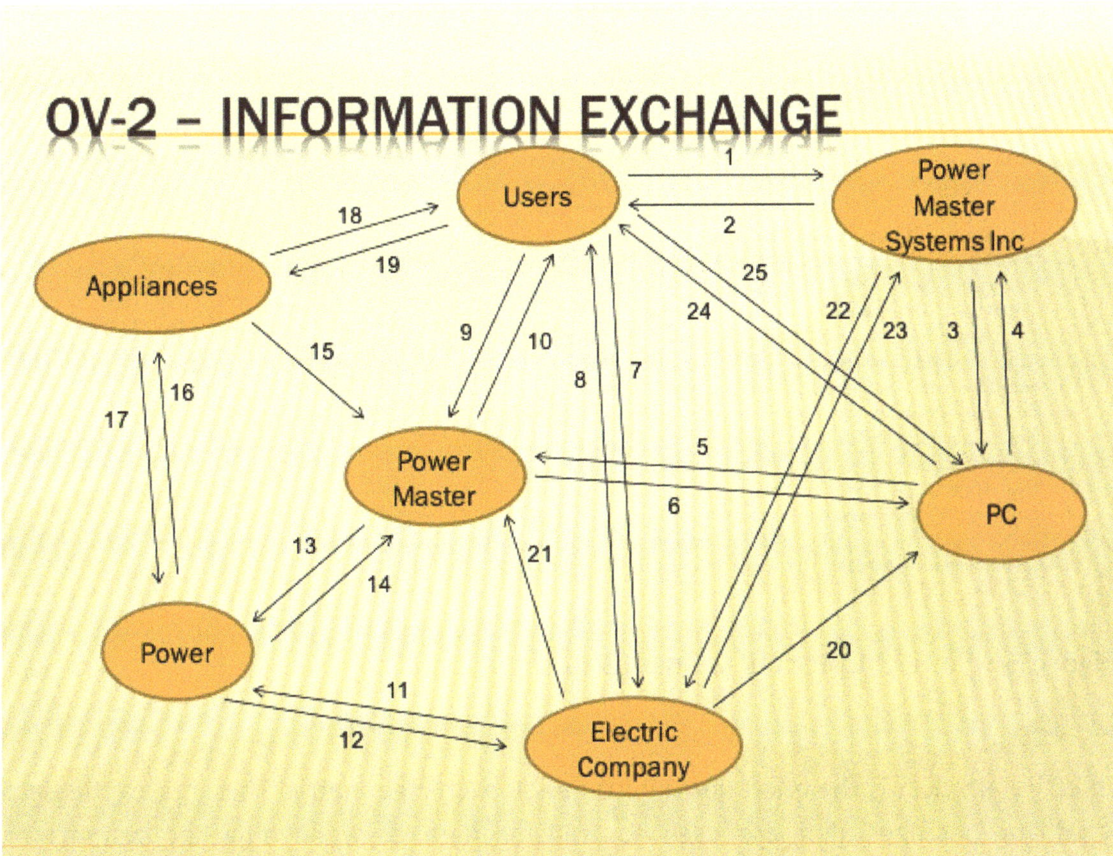

Figure 53: Power Master system OV-2

OV-2 – INFORMATION EXCHANGE

Need Lines:

1. Users interact with Power Master Systems Inc for support
2. Power Master Systems Inc will communicate new products, and communicate product recall and support information to users
3. Power Master Systems Inc will periodically provide software updates for GUI
4. Anonymous usage information can be collected from users who opt-in
5. PC will communicate with Power Master, can remotely turn off individual units.
6. Power Master software will relay power usage information to PC, where it can be saved or printed.
7. Users will pay their electric bill to the Electric Company. Hopefully the Power Master have helped reduce their bill significantly.
8. Electric company will send their bills with power consumption to the users.
9. Users plug in the Power Master physical unit inline between their appliance and an AC power outlet, and configure the device.
10. The Power Master system feeds power consumption data from their appliances back to the users.
11. The Electric Company can shut the house or business power off if bills aren't paid.
12. Power usage data is reported back to the electric company.
13. Power Master can cut the power of a device remotely.
14. Power usage is reported to the Power Master unit.
15. Power usage of appliance is calculated by the Power Master.
16. Power outlet checks to see if an appliance is plugged in. If it is, it provides electricity.
17. If appliance blows a fuse, the power outlet stops providing electricity.
18. Certain appliances have indicators indicating whether the unit is operating or not.
19. Users tell the appliances how to operate, and this can affect its power consumption.
20. The electric company provides cost information to the PC via the internet.
21. The electric company provides cost information to the Power Master software.
22. Power Master Systems Inc communicates with Electric Companies to get info, permission to use products, etc.
23. Electric companies communicate with Company with issues, concerns, cost info.
24. The PC can print off or save power consumption data from the Power Master for the users to see.
25. Users can interact with the PC to compile information they need from the Power Master.

Figure 54: Power Master OV-2 Data Elements Table

OV-3

The OV-3 product is the operational information exchange matrix. This creates the data words or packages that are used to transfer data between adjacent systems or subsystems in the development. This view is particularly important in data-intensive System Development efforts. It takes the form of a detailed table, including:

- **Table Headers Specified in Framework**:
  - ☐ Name of Operational Needline Supported (from OV-2)
  - ☐ Name of Information Exchange
  - ☐ Nature of Transaction (Mission/Scenario, Language, Content, Size/Units, Media, Collaborative or One-Way?)
  - ☐ Purpose or Triggering Event

- ☐ Information Source (ID of Producing Node Element, Owning Organization of Node, Name of Producing Activity, UJTL ID)
- ☐ Information Destination (ID of Receiving Node Element, Owning Organization of Node, Name of Receiving Activity, UJTL ID)
- ☐ Performance Requirements (Frequency, Timeliness, Throughput, Other)
- ☐ Information Assurance Attributes (Classification Restrictions, Criticality/Priority, Integrity Checks Required, Assured Authorization to Send/Receive)
- ☐ Threats (Physical, Electronic, Political/Economic)
- ☐ Operational Environment (Weather, Terrain, Policy/Doctrine Constraints)

OV-4

The OV-4 is the organizational construction that is used for the operation of the system. It provides an understanding of work responsibilities and task accountabilities for the system. An example is shown in Figure 55.

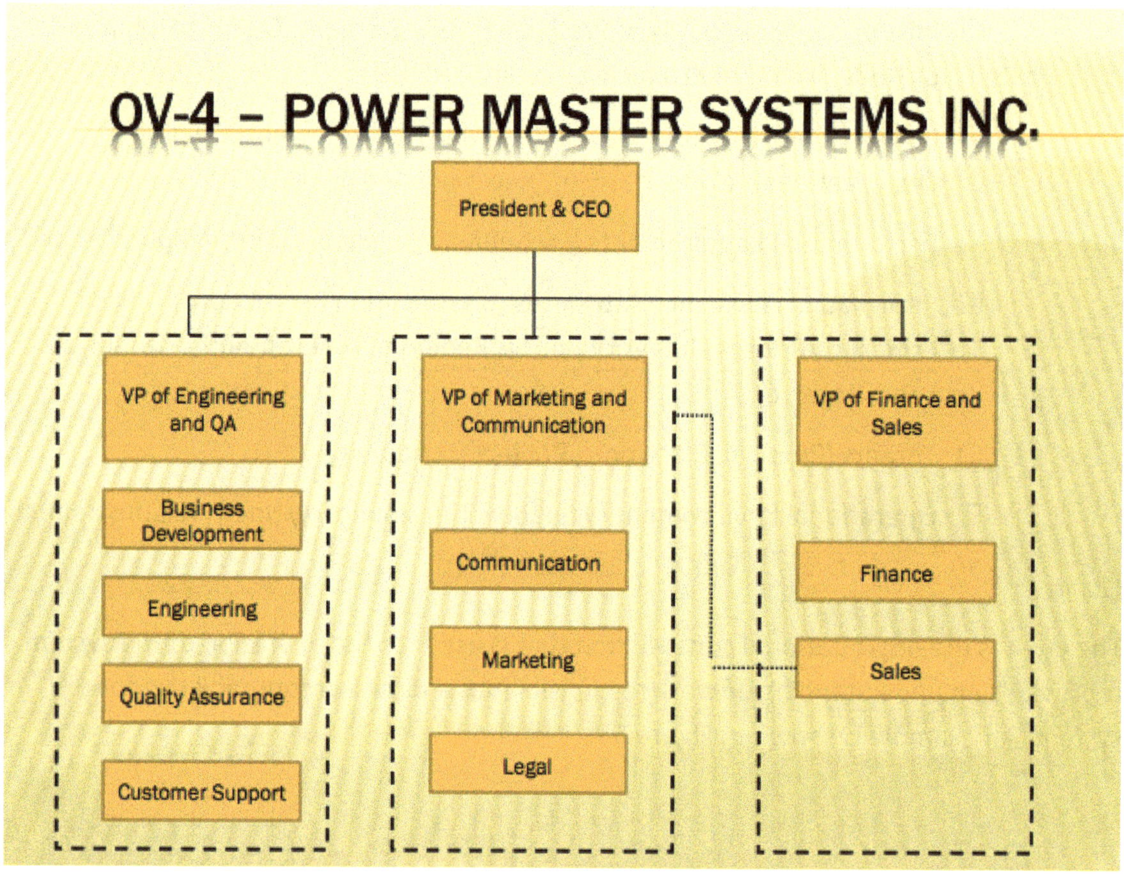

Figure 55: Power Master system OV-4

OV-5

The largest and most complex diagram in the framework is the OV-5, the operational activity model for the system to be developed.

We will begin by defining the various elements necessary for our toaster System Development, which includes Users, User Needs, System Requirements, States and Modes, and Detailed Requirements:

User Needs:

Toast Products	Quality	Timing
Safety	Efficiency	Maintainable
Size	Intuitive / User Friendly	
Adjustability	Aesthetics	

System Requirements:

120 VAC 50/60 hz	User Adjustment	Power Cord LG/Plug
UL listed	Indicators	Maintenance Access Panels
Portable	Safety Overrides GFCI	
Child Safe	Auto Shut-off	Art Deco / Chrome Finish / Pearl
Toasting Rate	Current Rate	Size / Wt

States:

- On
- Off
- Maintenance
- Storage
- Modes:
  - Toasting Bread
  - Toasting Bagel
  - Single Slot
  - Multiple Slot
  - Cooling
  - Adjustment Positions
  - Shutdown

Detailed System Requirements:

Number	Requirement	SuperReq	SubReq
1.0	**Usability**		
1.1	The system shall provide status indication		2.4
1.2	The system shall have an auto-shutoff	3.3	
1.3	The system shall be language independent	6.1	
2.0	**Operational**		
2.1	The system shall toast 1 product / minute		1.1
2.2	The system shall use 120 VAC		4.4
2.3	The system shall be portable		
2.4	The system shall allow user adjustment of toasting level	1.1	
3.0	**Security/Safety**		
3.1	The system shall use a GFCI	3.2	
3.2	The system shall be child-safe		3.1
3.3	The system shall fail to a power off condition		3.1
3.4	The system shall have a UL listing	3.2	
3.5	The system shall have a cord of 12 inches or less	3.2	
4.0	**Physical**		
4.1	The system shall have an art deco design		
4.2	The system shall have a chrome finish		
4.3	The system shall not exceed 12X12X12in		
4.4	The system shall use a standard LG Plug	2.2	
4.5	The system shall not provide hot surfaces to the user	3.2	
5.0	**Support**		
5.1	The system shall operate without maintenance for a period of three years		5.4
5.2	The system shall be replaceable, not repairable	5.2	
5.3	The system shall have maintenance panels	5.4	
5.4	The system shall require only cleaning maintenance beyond three years	5.1	5.3
6.0	**Training**		
6.1	The system shall require no training		6.1

Figure 56: Toaster System Development

The intention of establishing all of these factors early is to begin the development of the system utilizing the system V diagram. This begins with users, and user needs as we progress down the left side of the V. The intention of this methodology is to begin with a given challenge or problem, and then to decompose that problem and continue decomposing until we ultimately end up with a large set of system requirements. This would be the equivalent of starting at the top of the diagram in Figure 57, and decomposing a large number of requirements in the middle of the diagram. As we continue with detailed development we begin to address the individual requirements with architecture artifacts, and then we build these up into subsystems and systems until we create a singular solution for the problem at the bottom of the diagram.

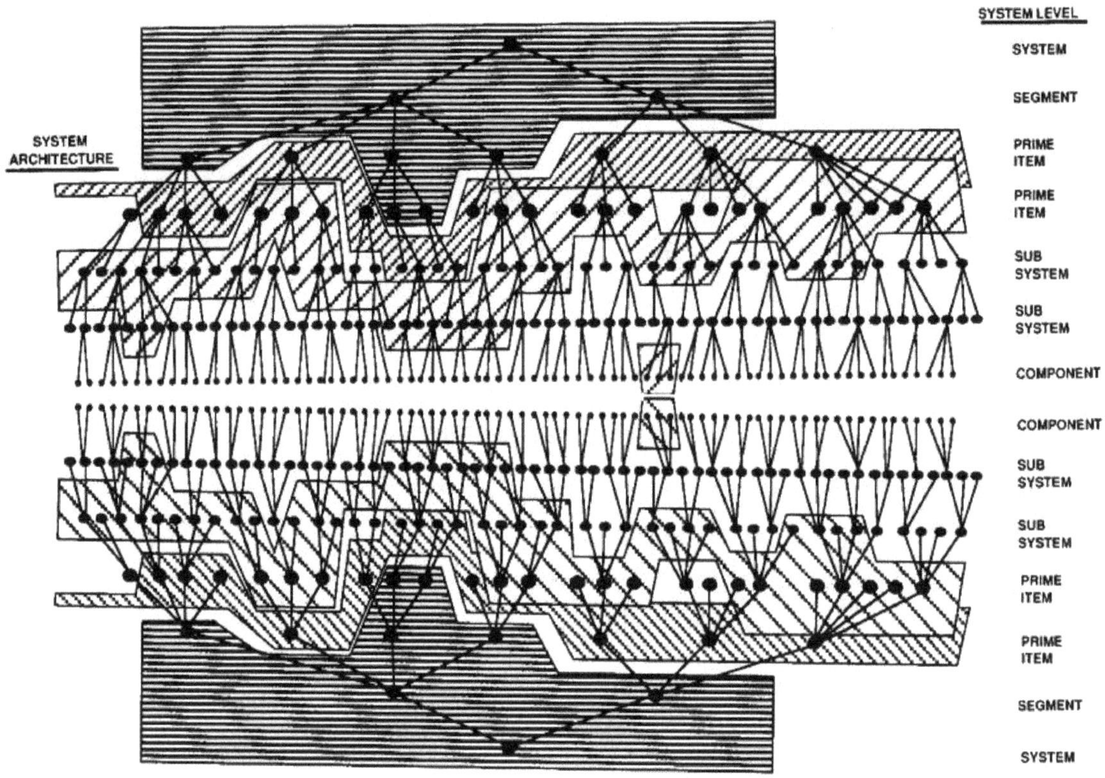

Figure 57: System Decomposition and Recomposition

The development of the OV-5, the Operational Activity Model is very similar to the Functional Analysis conducted in the Project Management section of this text. The principal difference between these two approaches in defining the operational activities is that the Functional Analysis is a bottom-up construction and the OV-5 is a top-down construction of the same Architecture. We will examine both methods to understand how they may be used in conjunction to produce more effective System Architecture than when used independently.

In our Functional Analysis, we made a list of functional or operational needs for the system. We then related those needs one to another, which established an operational flow diagram. This is shown in Figure 58.

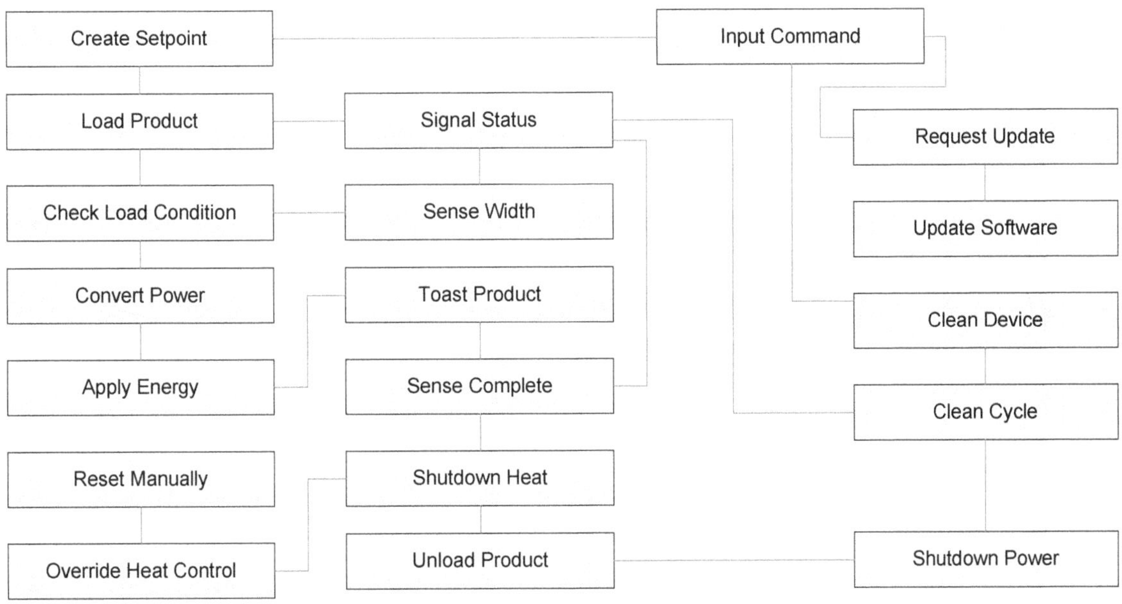

Figure 58: Toaster Functional Analysis

This exercise is of significant value, as it identified several functions not discerned at the outset of the project. It also clarified further assumptions that may have limited or expanded our thinking on the subject and reduced our aperture on the program by limiting capabilities or market segment. However, we have several more steps to go before we have established an effective System Architecture for our development.

To begin our OV-5, we will rely on the Integrated Definition for function modeling or IDEF0 modeling language. This is a federal information-processing standard based upon a variety of software and hardware modeling languages currently in use. IDEF0 has been used by a wide variety of commercial and government entities in establishing Architectures for complex system design.

IDEF0 was created to provide a means for completely and consistently modeling the activities, actions, processes, operations required by a system and the functional relationships relating information or objects that support the integration of those functions. It also provides a generic but rigorous, precise and reasonably concise modeling technique that is useful in a wide variety of System Development projects.

IDEFO is a method designed to model the decisions, actions, and activities of an organization or system. It was derived from a well-established graphical language; the structured analysis and design technique (SADT). The United States Air Force commissioned the developers of SADT to develop a function modeling method for analyzing and communicating the functional perspective of a system.

The principal characteristic of the IDEFO modeling language is the cell modeling graphic representation. The 'box and arrow' graphics of the diagram show the function as a box, and interfaces to or from the function as arrows entering or leaving the box. To express functions, boxes operate simultaneously with other boxes, with the interface arrows constraining when and how operations are triggered and controlled. The basic syntax for the model is shown in Figure 59.

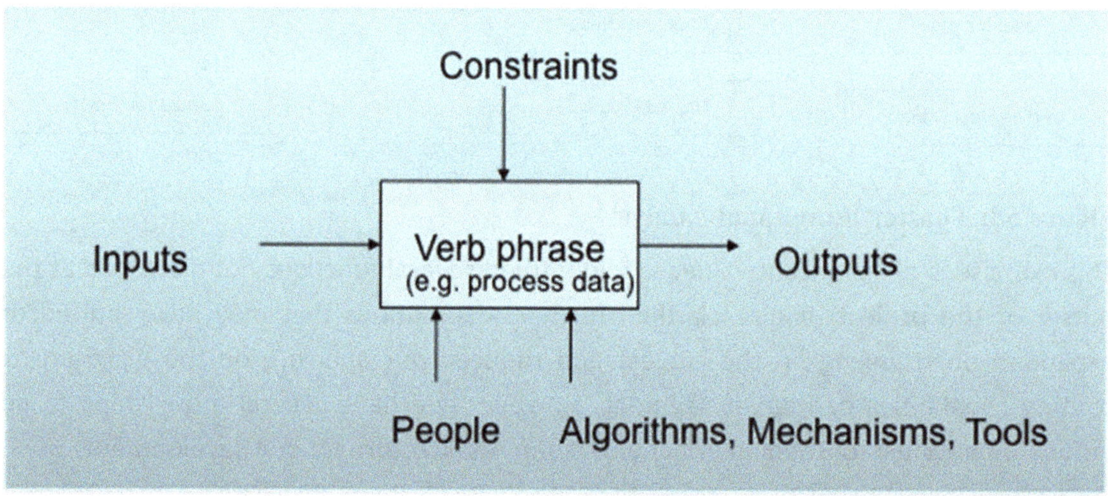

Figure 59: OV-5 construction

At the highest level, this diagram is known as a context diagram. In this view, the system is entirely encompassed by the boundaries of the box in the diagram. As we continue to add hierarchical structure to the diagram we 'Drill Down' into the box, identifying the functions and capabilities that make up the box and add levels of detail to the system.

Therefore, the context diagram is decomposed into related dataflow diagrams that can be recomposed into the whole. This is shown in Figures 60 and 61.

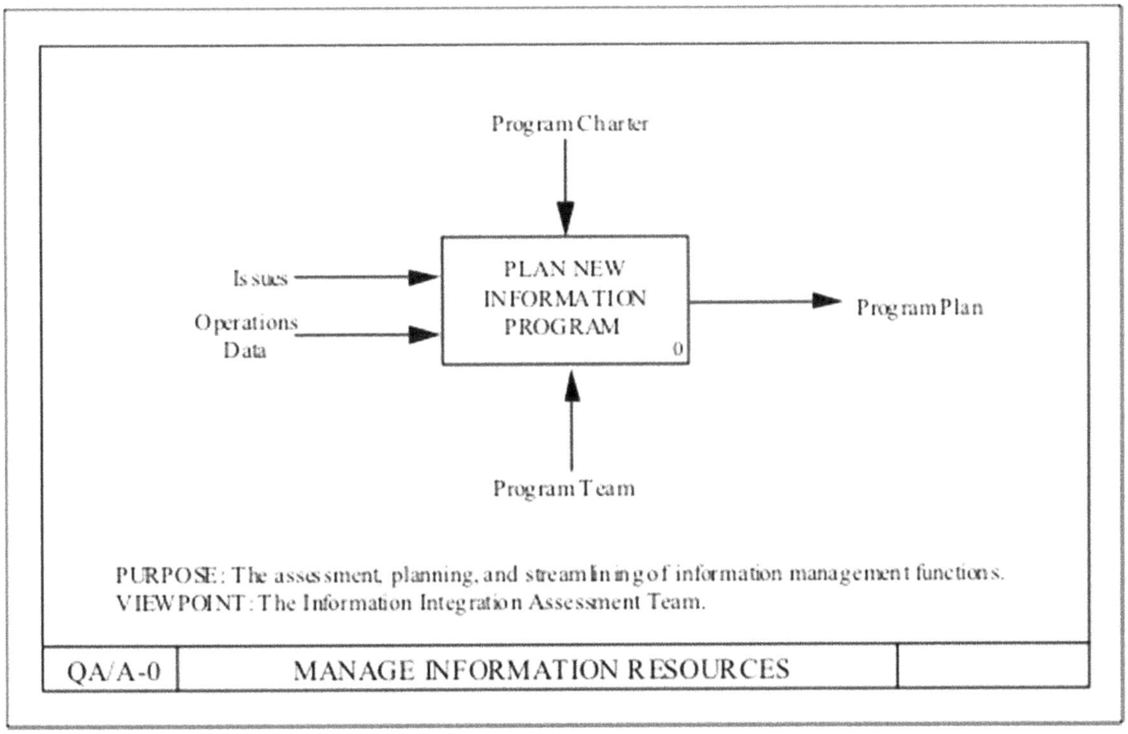

Figure 60: Context Diagram

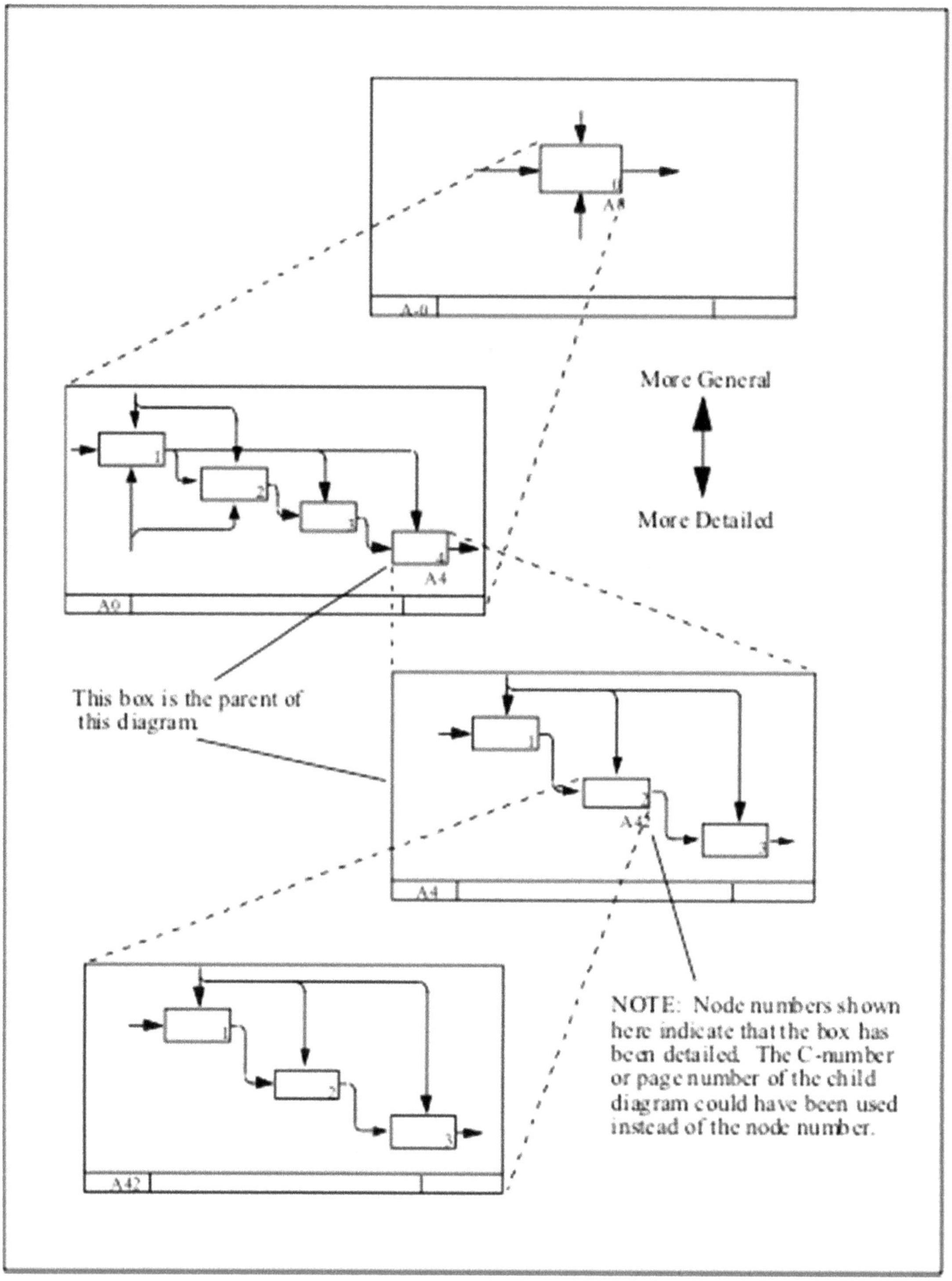

Figure 61: Decompositions of a System

These decompositions relate directly to our previous diagram and clarify how we are adding increasing levels of insight and detail to our system model, ultimately creating our System Architecture.

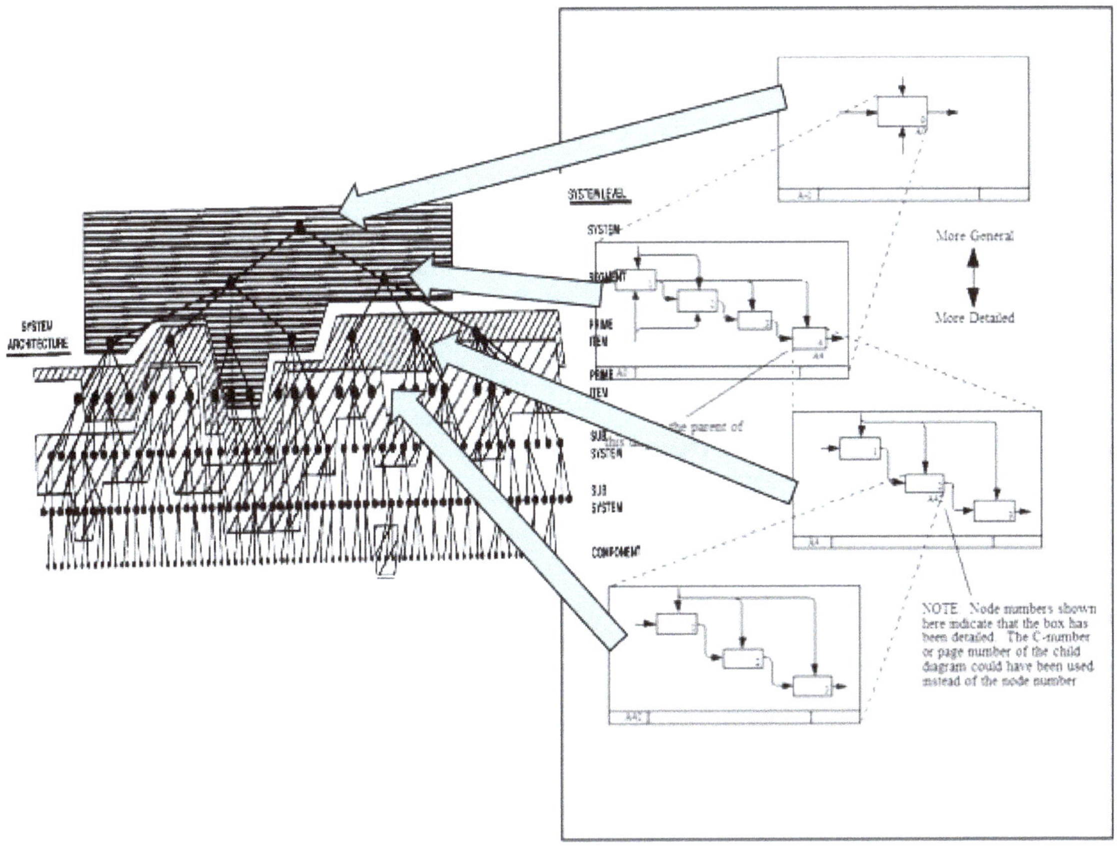

Figure 62: Decomposition leads to Requirements

This methodology allows us to take extremely complex and challenging systems and decompose them into manageable pieces until we create a System Architecture that may be addressed by Design Engineers. To illustrate this fact we will take a simple example: <u>getting an individual to work in the morning</u>.

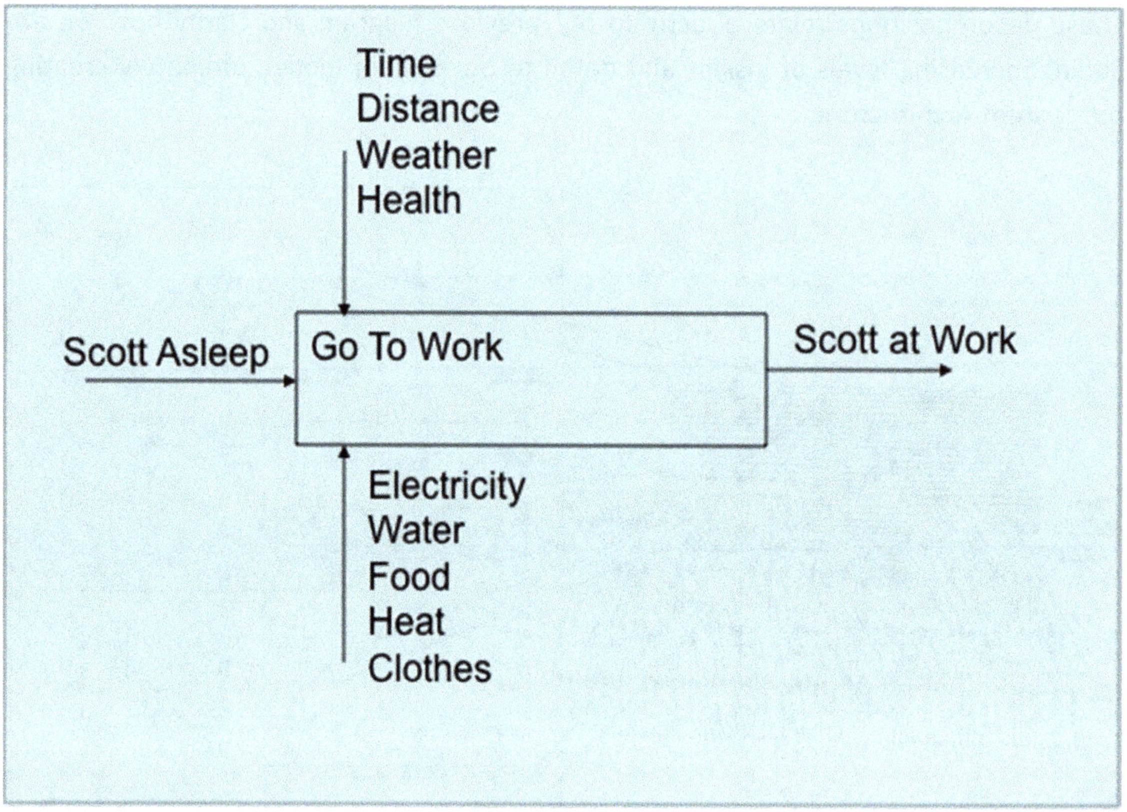

Figure 63: Context Diagram

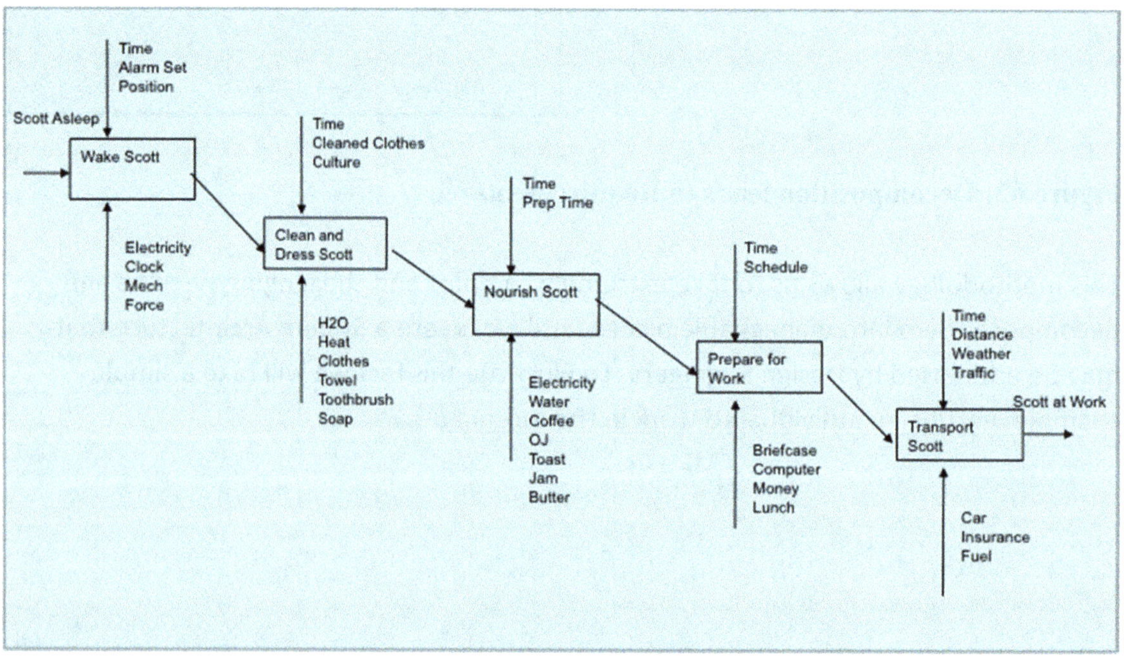

Figure 64: Drill Down 1

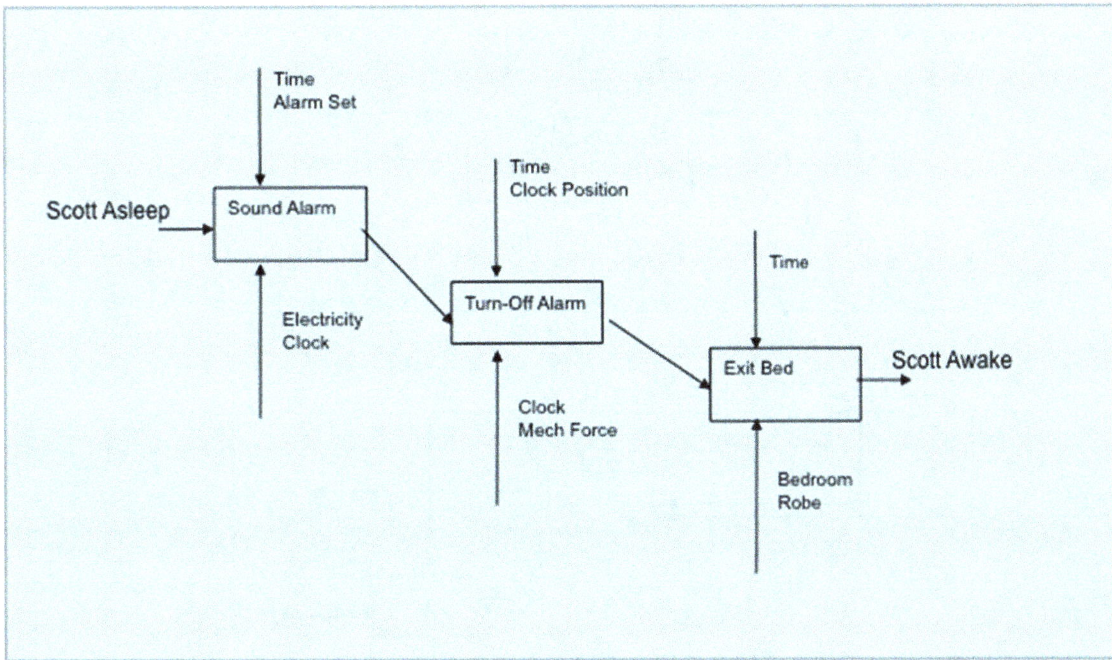

Figure 65: Drill Down 2

As can be seen, each Drill Down further elaborates on the functions within the box that it represents. Each must maintain referential integrity with the parent diagram, as shown in Figure 66.

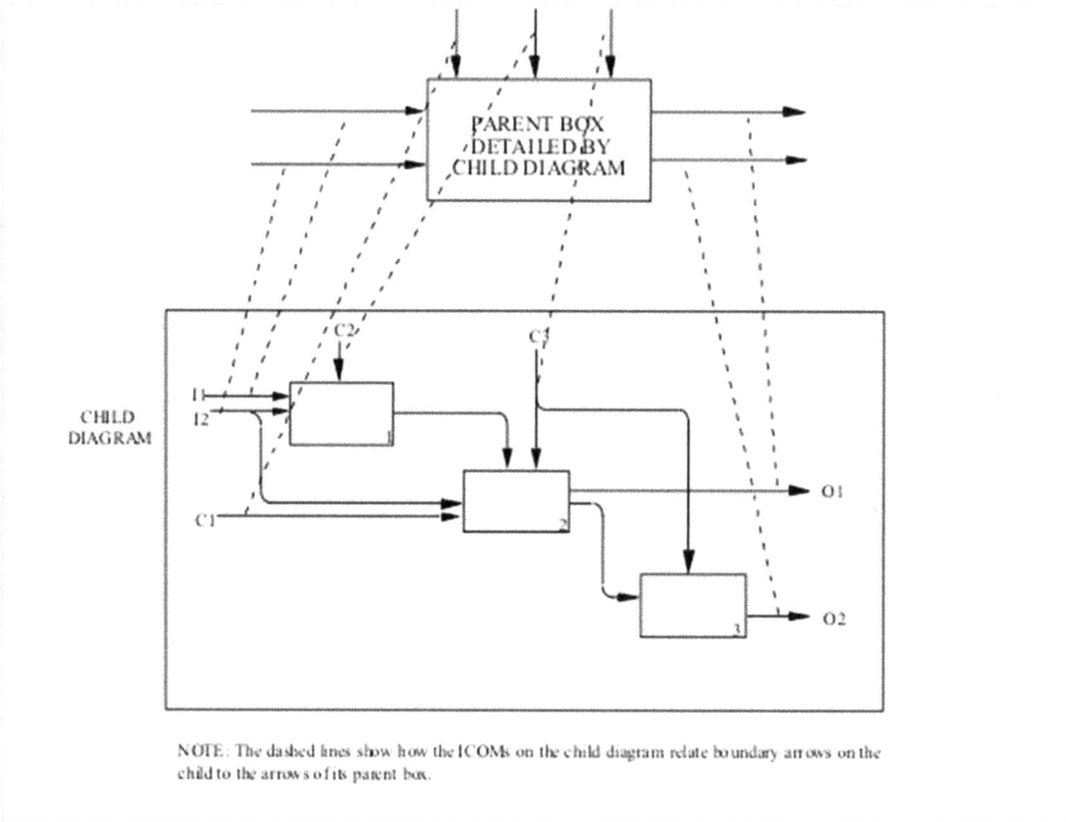

Figure 66: Referential Integrity of diagrams

This effort can be continued until all of the boxes in the system are well-understood or purchased components. We will now examine a more complex problem, one of performing a caesarean section in a hospital.

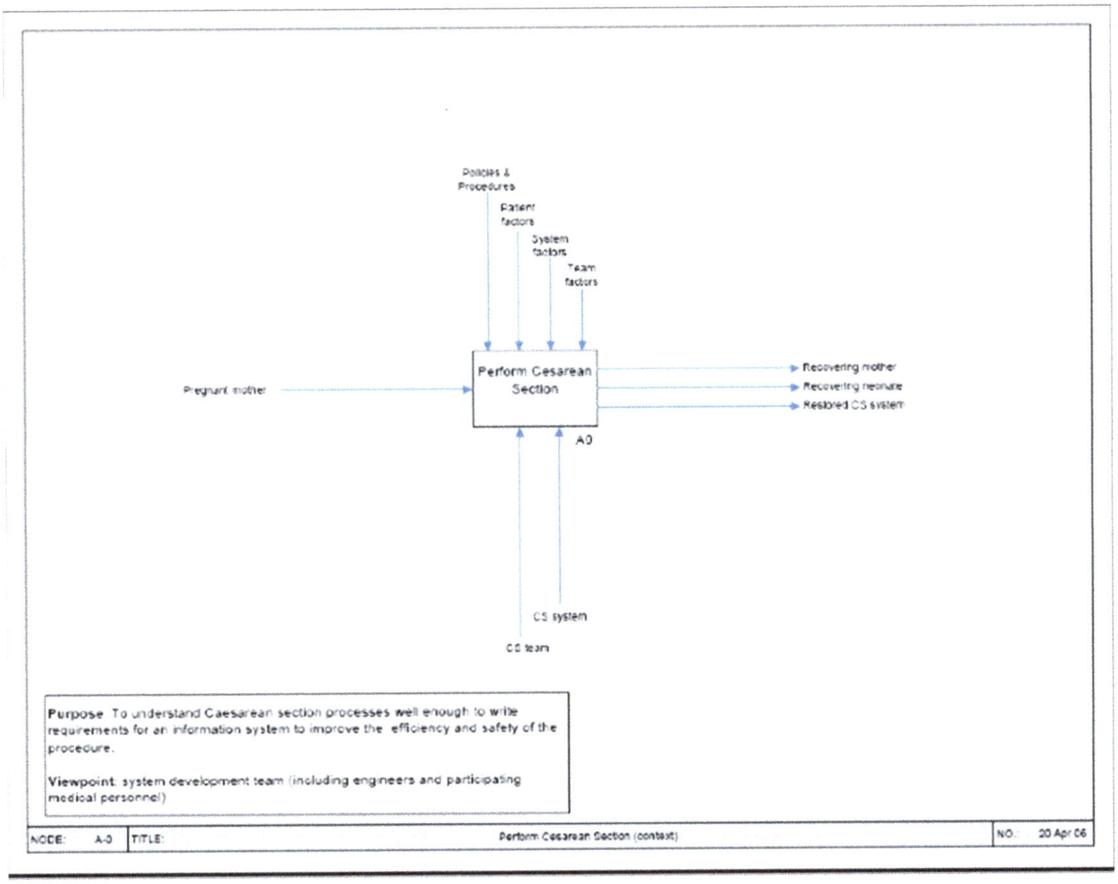

Figure 67: Context diagram CS

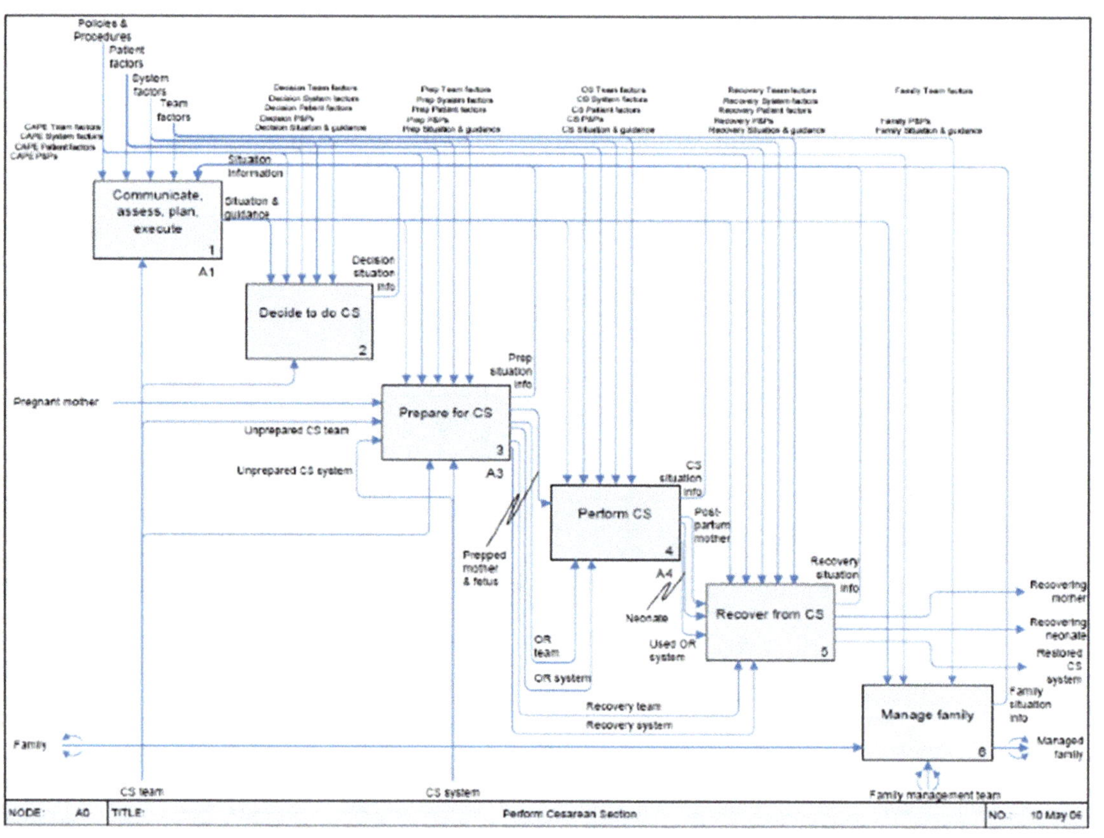

Figure 68: First Drill Down CS

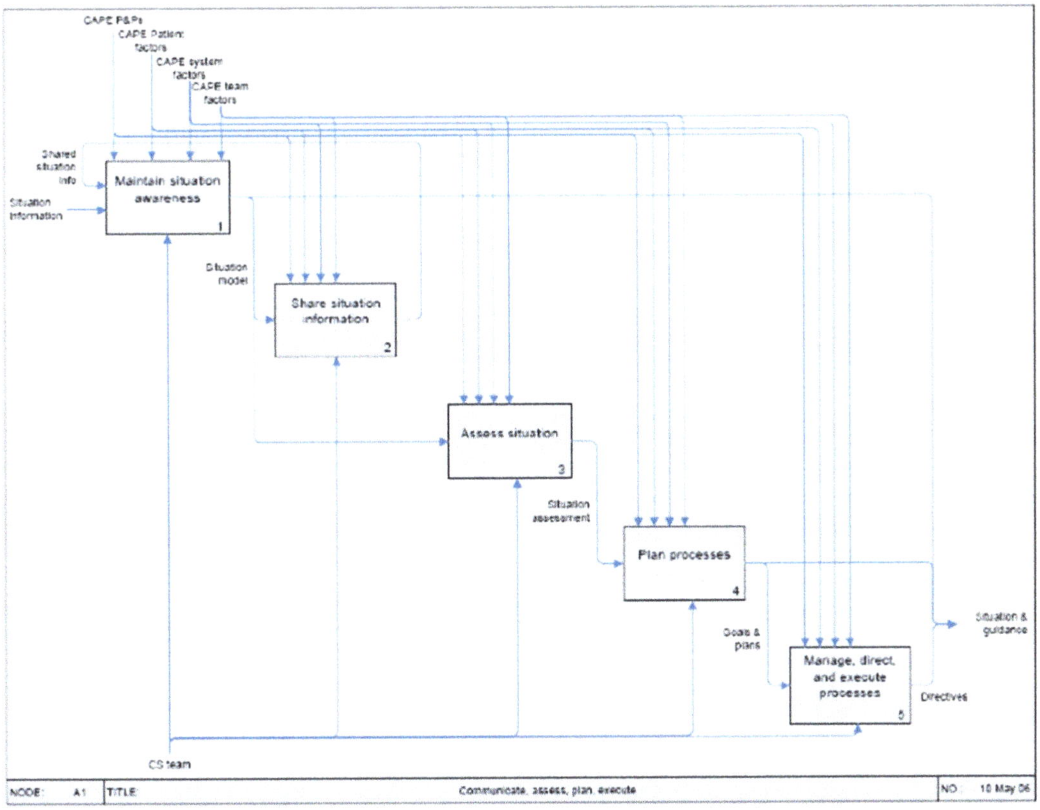

Figure 69: Second Drill Down CS

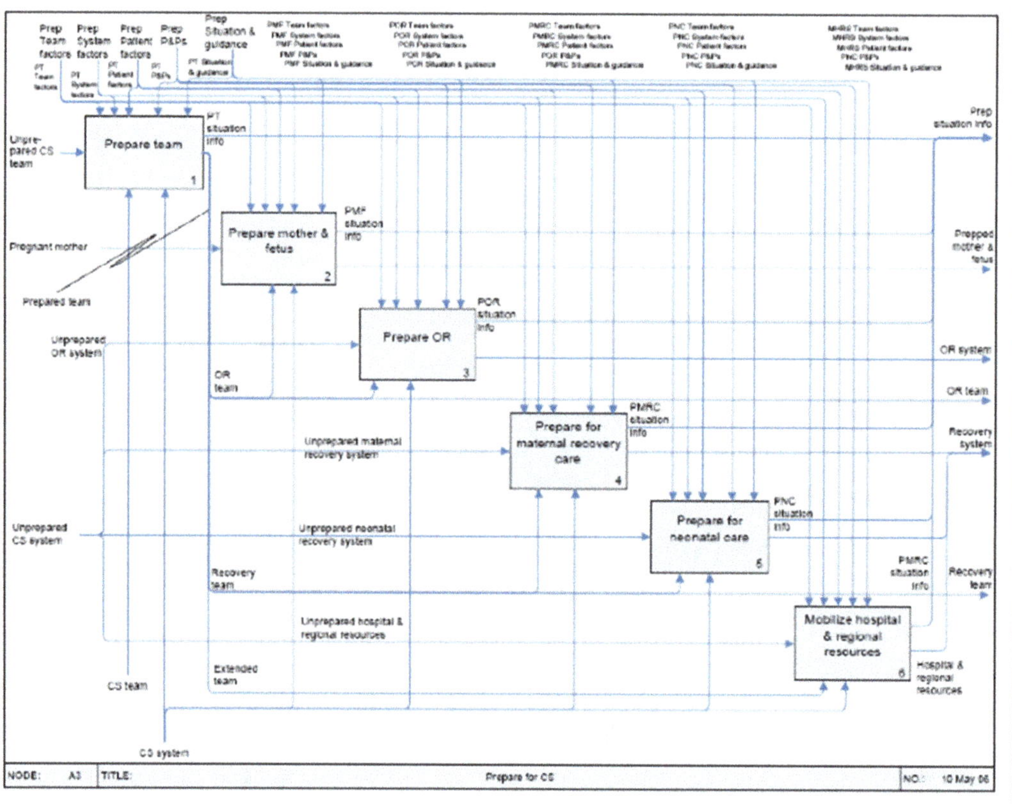

Figure 70: Second Drill Down CS

This example provides a great deal of detail associated with the complexity of performing a cesarean section in the hospital. Please note that these diagrams should contain an appropriately numbered title block, and should demonstrate referential integrity. Careful adherence to the rules and methods of producing these diagrams can provide an extremely straightforward and easy communication tool for presenting operational concepts to other members of the development team.

As a final example, we will look at a very simple device, demonstrating that these techniques of development for a System Architecture are applicable in a wide variety of domains, and are not limited only to very complex situations. In this case, we have chosen to diagram the operations of a small toy, a water pistol. This device is shown in Figure 71.

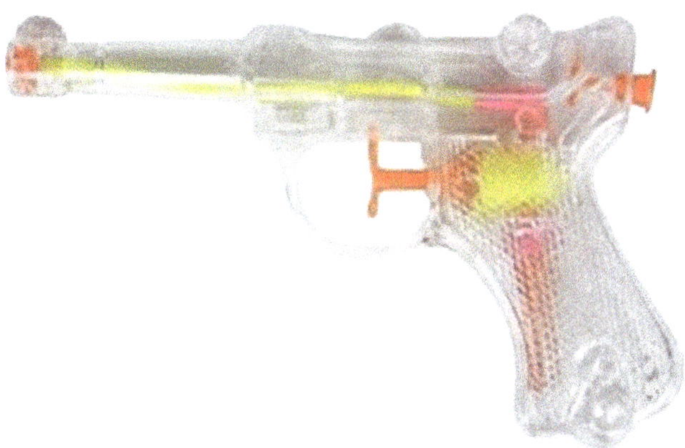

Figure 71: Water Pistol System

A simplified diagram is also included to clarify the operation of this system. The subsequent System Architecture being developed is based upon this simplified operational diagram.

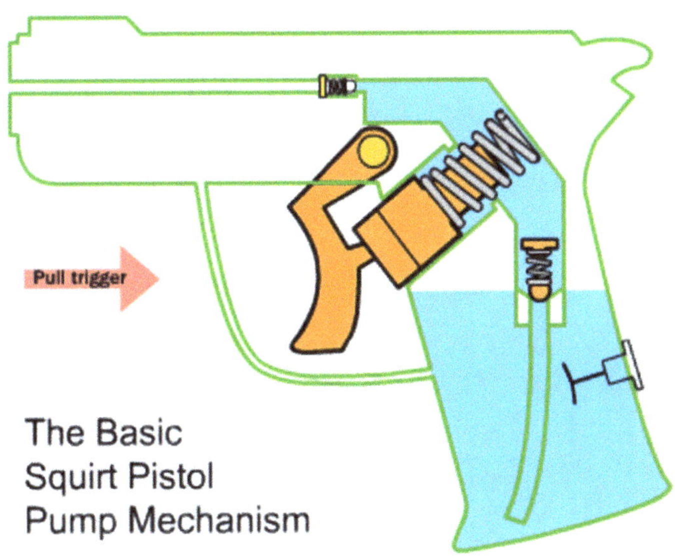

Figure 72: Operational Concept of Water Pistol

As we have previously done, we will first create a Functional Analysis of this system with an included consolidation of functions to provide a type of System Architecture for this device. This is shown in Figure 73.

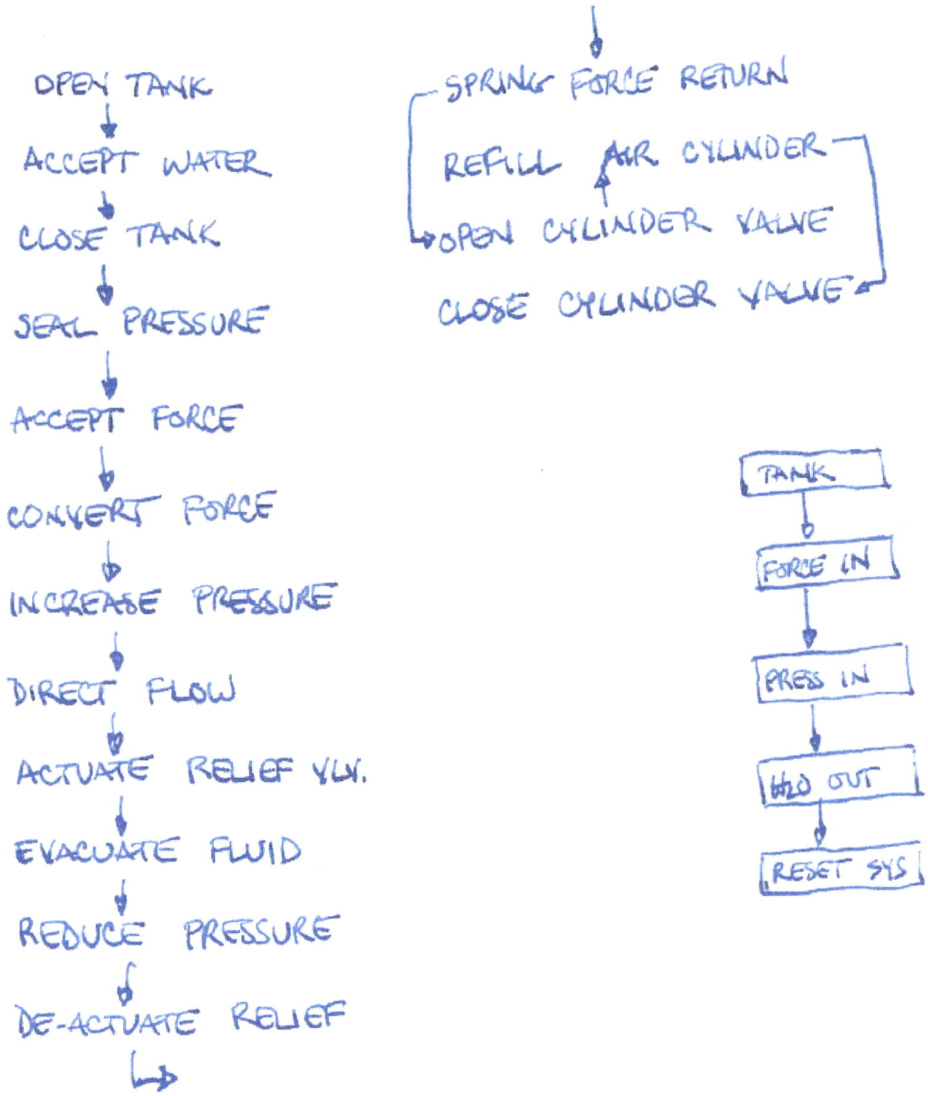

Figure 73: Functional Analysis of Water Pistol

We will also utilize our previously discussed techniques to develop the OV-5 in order to ascertain whether our Functional Analysis is complete and whether our System Architecture is fully prepared. The context diagram and several Drill Downs are included in the following diagrams.

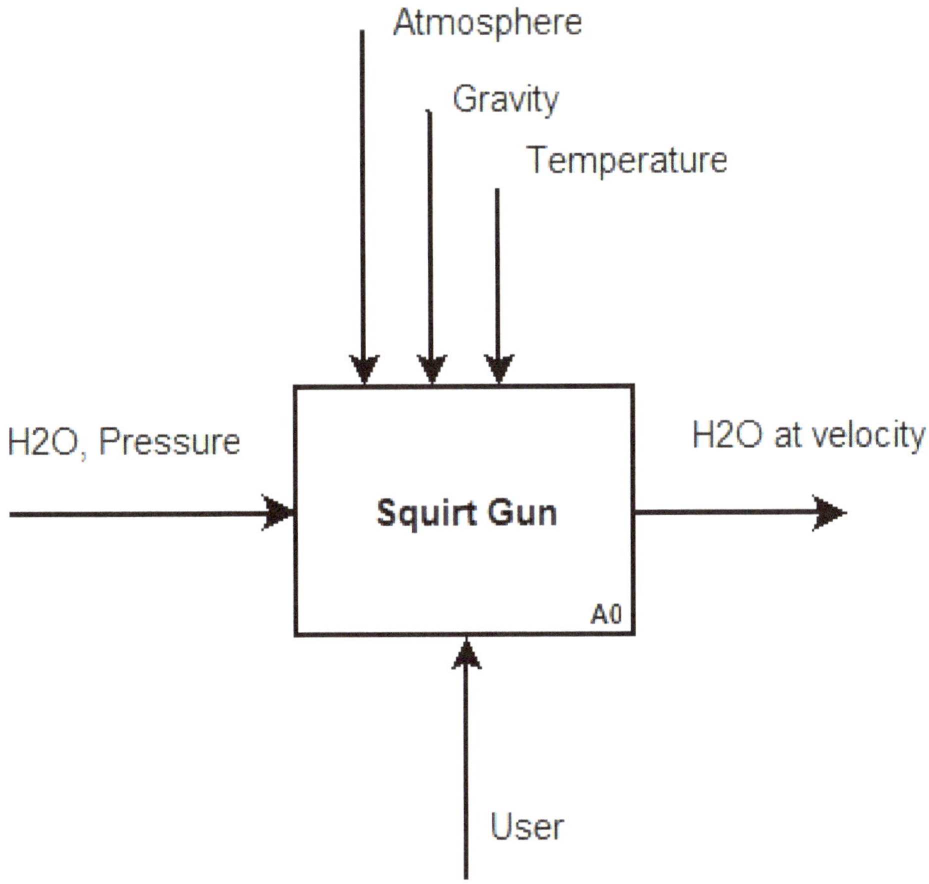

Figure 74: Context Diagram for water pistol

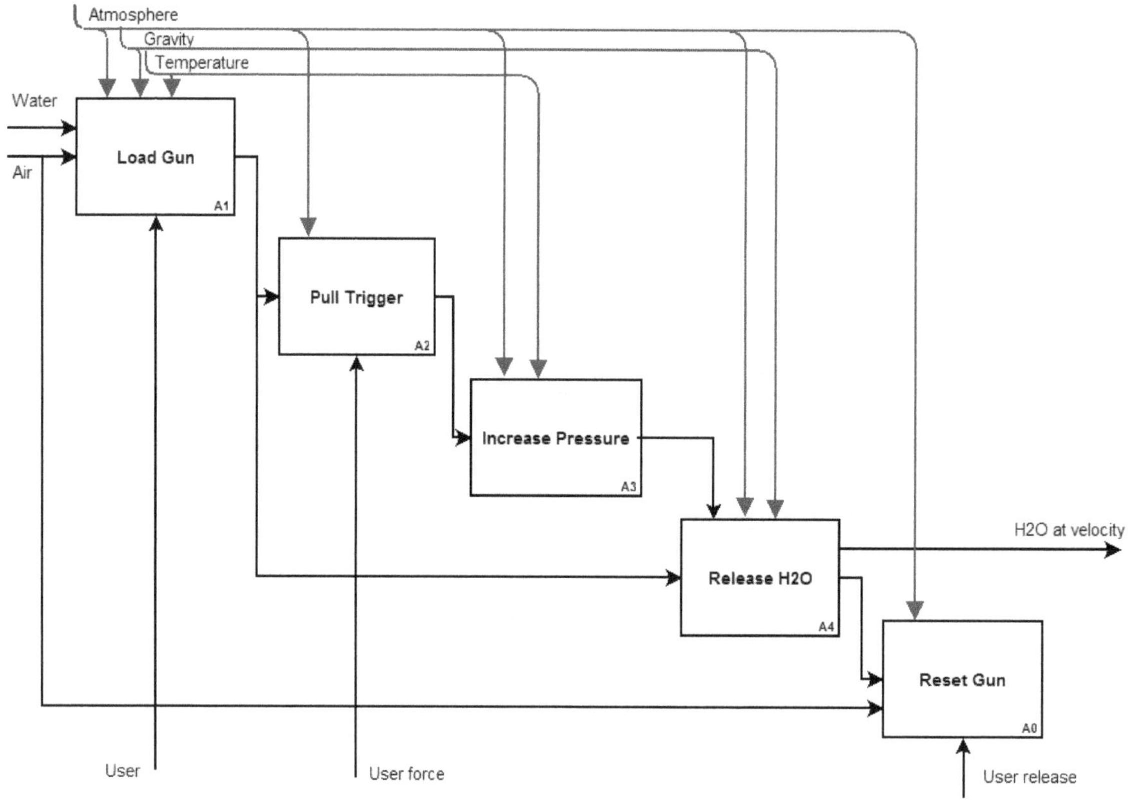

Figure 75: First Drill Down of Water Pistol

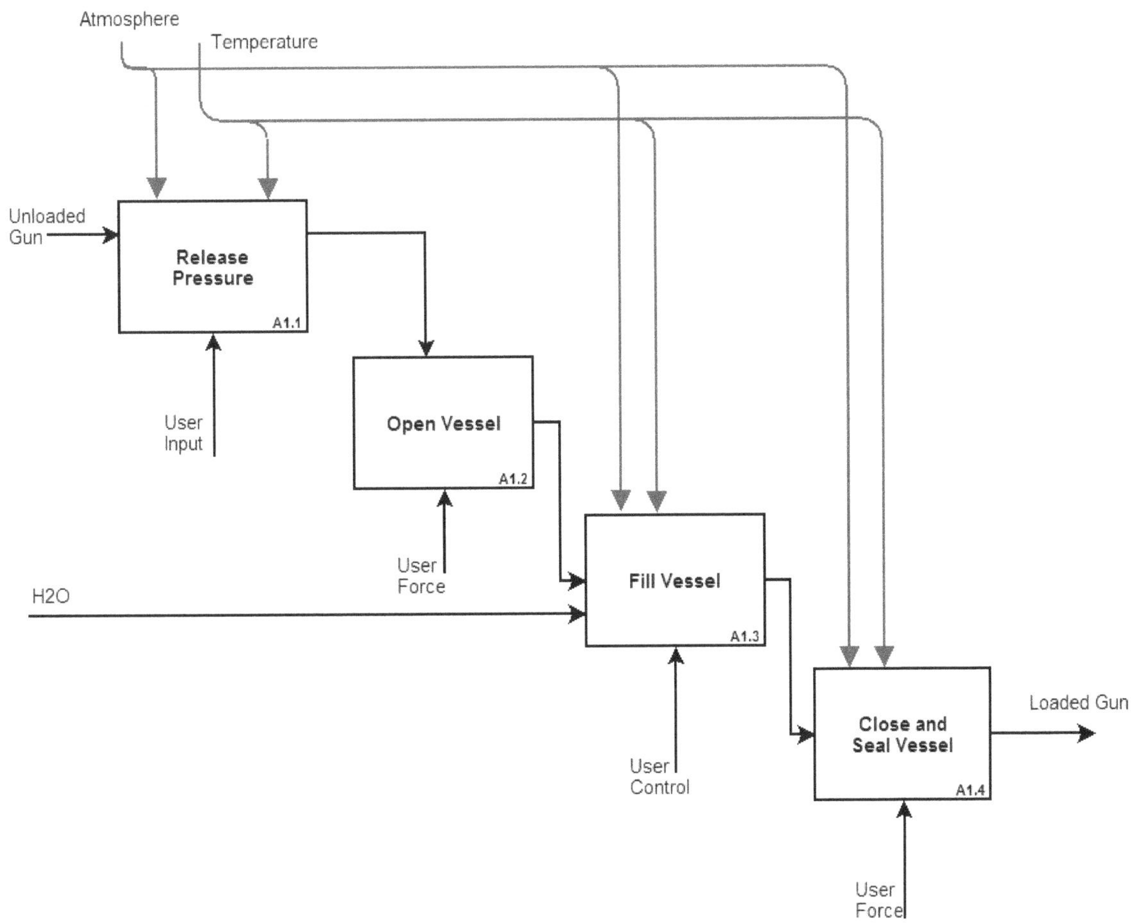

Figure 76: Second Drill Down of Water Pistol

The first Drill Down is remarkably similar to the System Architecture defined by the Functional Analysis. Had we completed all second Drill Down diagrams, we would have seen that each of the elements noted in the Functional Analysis appear in some form in the System Architecture on the OV-5. Utilizing these two techniques simultaneously enables us to better ensure that we have a complete description of the operational activities of the system.

Completion of the OV-5 establishes the System Architecture for the development effort. As might be expected, insights gained and conclusions developed through the generation of these work products often require us to go back and modify or rewrite previously defined work products for the system. In the end, we would like all our documents to be consistent so that they might be used as a record for future system developers who may be faced with a similar challenge, and as a justification of decisions made during the development effort.

Once the System Architecture is completed, it can be parsed out to the various Mechanical, Electrical, and Software Engineers who will be called upon to complete the Physical Architecture for the system. The System Architecture provides a generic operational diagram, and as long as the Component Engineers adhere to the logical constructs defined in the System Architecture, the manner in which the functions within the box are completed is of little concern to the Systems Engineer. Thus the Component Engineers are 'white-box' developers, in that they are concerned with what is inside each box and all of the details of the functionality. The Systems Engineers, on the other hand, are 'black-box' developers, as they are less concerned with the actual functionality that occurs within the box as long as it adheres to the logical construction defined in the System Architecture. This is one of the primary reasons why a Systems Engineer does not need to be an expert in software development or hydraulics in order to design a system containing those elements.

This is not to say that it is useless to have specific Engineering skills in a Systems Engineering endeavor. On the contrary, those skills are extremely useful, but when they are applied the Systems Engineer is acting in the role of a Component Engineer. As might be expected, the line drawn between these two domain areas is a gray area, and can often be quite difficult to define. Suffice it to say that the Systems Engineers and Component Engineers work hand-in-hand to assure the completion of an effective and coherent system.

The Development of Requirements

One of the largest challenges in accomplishing System Development effectively is defining the problem to be solved and the criteria that will be used to establish if it has been solved adequately. Once we have defined our list of requirements, we have defined the criteria that will be used to assess the quality of the solution and its ability to meet the customer's needs.

Many engineering organizations maintain that a handful of requirements are adequate to complete any development project. While it is true that more requirements are typically more expensive to accomplish in the development process, the number of requirements is related to the quality of fit of solution. In other words, if adequate time is taken in the preparation and development of requirements, the validation of the system will be much more readily accomplished and the customer will be more satisfied.

Taking the example of our toaster, one might ask what an adequate number of requirements might be. If a rigorous requirements development is pursued with regard to this project, we might expect that there would be two hundred to three hundred requirements developed. If we were to consider a large-scale project such as the installation of a stadium or the development of a new tactical aircraft, we might expect that list to be much more extensive, perhaps twenty thousand requirements or more.

Of course it is unlikely that any requirements development group can simply sit down and begin writing a large number of requirements, and it is even less likely that those requirements will be evenly distributed across the needs of the program. Therefore, we have spent significant time developing a System Architecture, which will be used ultimately to develop the majority of the operational requirements for the system.

As previously stated, when we examine any work product for a system, we will observe many functional descriptions and need lines associated with those products. Every functional box and every need line will ultimately produce at least one requirement, and potentially more than one. This is likely to generate large numbers of requirements that would not otherwise be readily apparent to the development team.

Requirements Engineering is the process of establishing the services that the customer requires from a system and the constraints under which it will operate after development. Requirements may be functional or non-functional, where functional requirements describe system services or functions and non-functional requirements typically result in a constraint on the system or development process.

Most large-scale System Developments deal with problems that are so complex that they will never be fully understood until we have undergone a significant portion of the system's development. Insight is often achieved by means of the decomposition process, but this implies that requirements will normally be both incomplete and inconsistent. Of course, we seek to minimize these two characteristics in our Requirements Development Process.

This includes:

Requirements Elicitation

Requirements Analysis
Requirements Validation
Requirements Management

As we have previously stated, Systems Engineers are responsible for the development of requirements throughout the decomposition process. Responsibility for program development is noted in the following diagram.

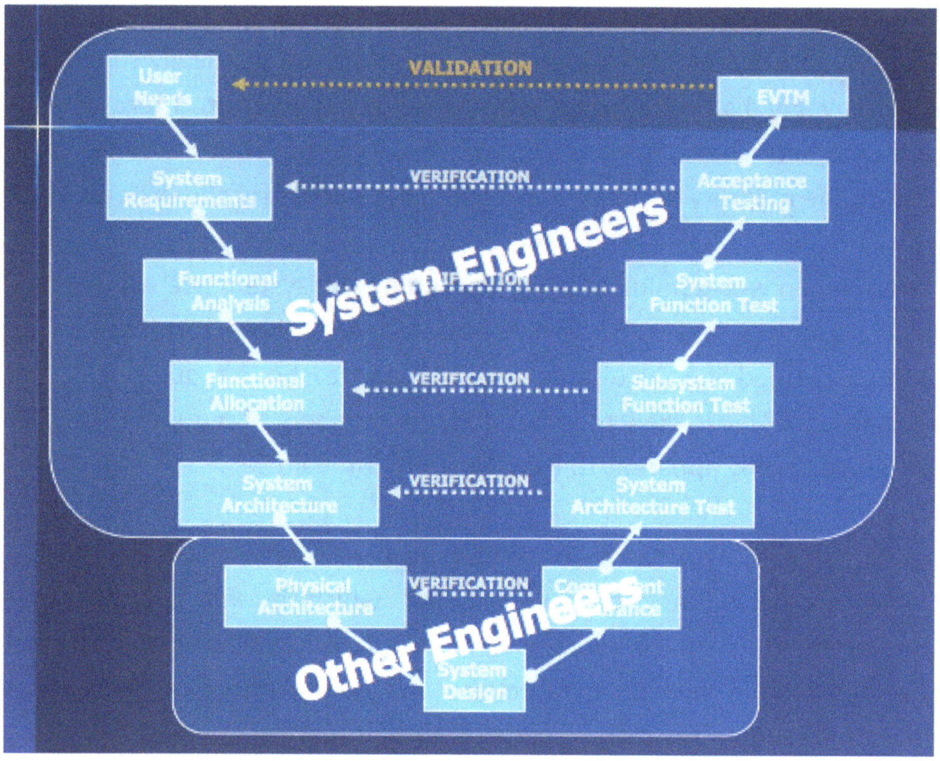

Figure 77: Development Responsibility

Although the responsibility for System Development will be passed from the System Engineer to the Component Engineers at the transition to Physical Architecture, requirements will typically be dynamic and potentially volatile until we accomplish and complete the detailed design of the system. This is largely due to the advances in knowledge that are created during the development process, and should not be considered a bad thing in the development. Minimization of this volatility will be of benefit to the organization with regard to the efficiency of resource use and waste.

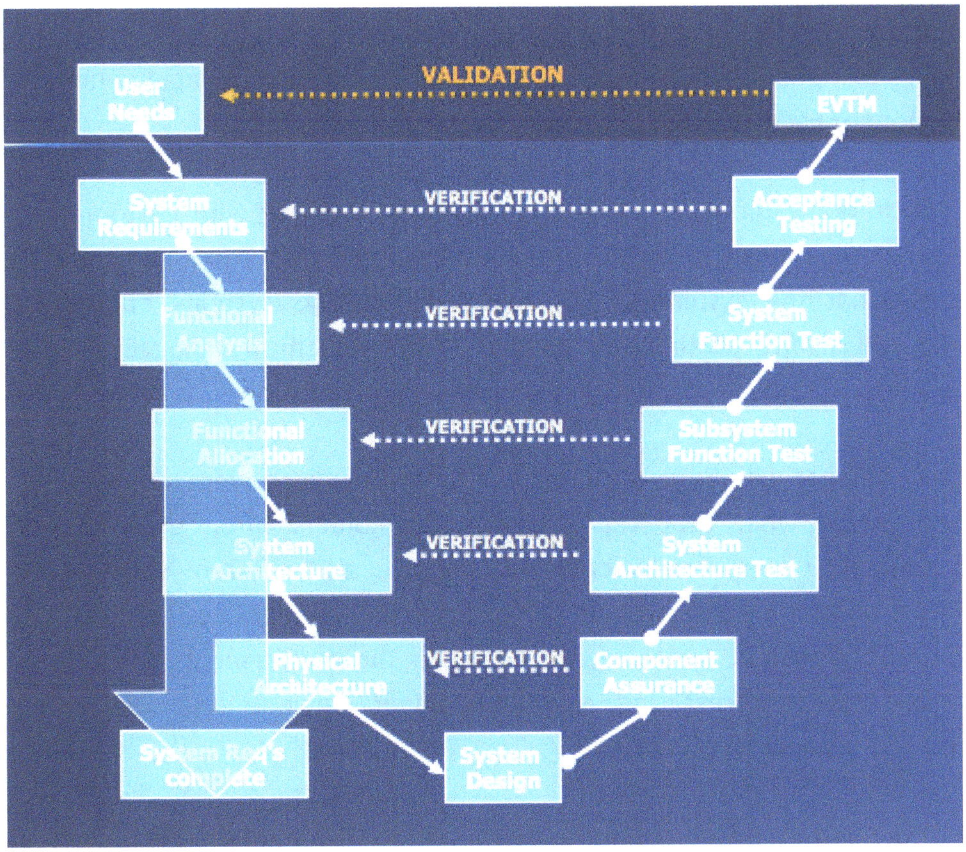

Figure 78: When are Requirements Complete?

In essence, during Requirements Definition we increase the aggregation of our knowledge of the system in order to fully decompose the problem. By virtue of that, we hope to expose all of the details of the problem to our insight and understanding. We follow a simple process to advance the resolution of our understanding:

Requirements definition–
A statement in natural language plus diagrams of the services the system provides and its operational constraints. This is typically a document written for customers.

Requirements specification–
A structured document setting out detailed descriptions of the system services. Written as a contract between a client and contractor, and typically used to bridge the gap between the System Architecture and Physical Architecture.

Design specification–
A detailed design description that can serve as the basis for a design or implementation; written for Component Engineers and Developers.

Requirements Elicitation typically involves technical staff working with customers to discern more about the application domain, the services expected and the customer's requirements of what the system shall provide, as well as systems operational constraints. This will involve all the stakeholders of the system including End-users, Managers, Engineers, Maintenance, Domain Experts, and Trade Unions.

As any System Developer will discover during this phase of the project, stakeholders typically do not know what they really want. Stakeholders often express their requirements using their own terminology and language, and different stakeholders have potentially conflicting requirements. Organizational and political factors will influence many of the system requirements. Many of the requirements will change during the development process and new stakeholders may emerge or existing stakeholders may be removed.

The intention of the System Development team during the Requirements Development Phase is to explore the many viewpoints of the stakeholders of the system. These different viewpoints are examined within the structuring of the requirements to represent the perspectives of different groups of stakeholders. This multi—perspective analysis is important as there is no single correct manner in which to elicit and analyze system requirements, and any known requirement is of benefit.

Many techniques maybe used to better understand the system, and these tasks may be utilized both prior to and after the development of many of the work products in our DoDAF approach. They include:

Interviewing – Used to gain insight into the stakeholders' perspectives on the system; although they are not particularly good for understanding the system's domain requirements.

Scenarios – Real-life examples of how a system may be used; they typically include a description of the starting situation, the normal flow of events, what may go wrong, and a final state when the scenario is completed. This analysis can be very useful in the development of operational activities and other products.

Use cases – A scenario-based technique used to identify the actors in an interaction and the system operation with regard to the actors. Sequence diagrams are used to add detail to use cases showing the sequence of events in the system.

Ethnography – Social scientists spend a considerable time observing and analyzing how people actually work. In many cases people do not have the inclination or knowledge to be able to explain or articulate how they accomplish their tasks. Ethnographic studies have shown that work is usually richer and more complex than that suggested and simplified system models. These studies can help to add significant detail to vague or abstract areas in the System Architecture.

Modeling and simulation – An abstract representations of a view of the system; a simulation is that model examined over time. Use of models and simulations often provide significant insight to the System Developers and define requirements and constraints that are not readily understood at the outset of the System Development. In fact, the development of our DoDAF work products is simply a form of modeling.

There are several guidelines for writing good requirements:
1) Requirements should contain a single thought
2) Requirements use the word 'shall' in the definition
3) Requirements should form a complete sentence

As requirements are developed, they will be incorporated into a Requirements Verification Traceability Matrix (RVTM). This matrix is designed to allow Requirements Management to be conducted. There are several elements of note to incorporate into this particular matrix:

1) The requirement statement must be adherent to our guidelines
2) The requirement must be verifiable
3) The requirement list must be consistent and not contradictory

Therefore, the RVTM is defined by the use of several columns of interest:
1) A number for each requirement
2) The description of the requirement in its entirety
3) The traceability for the requirements (i.e. where it came from), and the verification method.

An example RVTM is included in Figure 79.

RVTM for Toaster Product

					Verification Method			
Requirement	**PBSS**	**OV-2**	**OV-5**	**Requirement**	**Inspection**	**Model**	**Test**	**Demonstrate**
1	1			The system shall operate on 120 VAC	X			
2	2			The system shall provide a manual shutoff	X			
3	3			The system shall not exceed 12X12X12 inches	X		X	
4	4			The system shall not mar countertops if slid on surface	X			
5	5			The system shall provide an indicator for operational status	X		X	
6	6			The system shall provide an audible alarm in the event of emergency shutdown	X			
7	7			The system shall autoshutdown in the event of overtemperature	X		X	
8	8			The system shall autoshutdown in the event of current draw over 10A	X	X		
9	9			The system shall weigh less than 4 lbs	X			
10	10			The system shall toast bread sices				X
11	11			The system shall toast bagel slices				X
12		1		The system shall allow user adjustment of setpoint				X
13		1		The system shall indicate setpoint for user				X
14		1		The system shall utilize an LG plug for power input				X
15		1		The system shall provide a push button start for the user				X
16		1		The system shall provide automatic unloading of product after completion of toast cycle				X
17		1		The system shall accept user loading of product				X
18		2		The system shall unload toast to an area at least 2 inches from heated surfaces				X
19		2		The system shall provide user audible notification of unloading event				X

Figure 79: Toaster RVTM

Collecting What We Know About the Problem

The vision of the project is continuously refined and expanded as the Systems Engineering development process progresses. This vision is communicated to others through the defined System Architecture, the Requirements List and the RVTM. If we accomplish our tasks appropriately, this information is adequate to complete the System Development. This System Architecture may then be handed off to the Component Engineers for completion of the Physical Architecture and detailed design of the system.

However, in the midst of the System Development cycle things are often discovered that question our original assumptions, constraints, or ability to successfully complete the project. Due to these shifts, scope changes are often necessary and appropriate in the development of the project. Areas of knowledge that are undefined or unknown at the time are also identified. Part of the effort of the Systems Development Team is to identify all of the information known about the system being developed, as well as the unknown areas where data does not exist to support any conclusions.

Defining What We Do Not Know About the Problem

When documentation related to what we know about the system is complete, a Gap Analysis may be conducted to clarify and define areas of uncertainty or unknowns on the project. This Gap Analysis is normally carried out by peripheral members of the development team, as those closest to the development are often clouded in their vision and may have a great deal of difficulty identifying gaps of knowledge or uncertainties.

A number of methodologies may be used to conduct the Gap Analysis. One such recommended method is a review of the System Architecture that examines each capability to be delivered with regard to the understanding of the System Architecture and logical construction of the capability. For example, in our water pistol example, an examination of the System Architecture may reveal uncertainty about the optimal placement for the relief valve that is used to release the water stream under pressure for the gun. A position might be defined based on personal preference or by identification of a model utilized to address this gap in knowledge. For example, two prototype water pistols with varying positions for the relief valve might be created, with each being tested for performance characteristics and the optimal position being defined by the test results.

In a different example, if we were asked to create a system to read the license plates of all cars passing over a given bridge in our locality, we may have very little idea how much data would be compiled in a twenty-four-hour period. We may choose to clarify this gap in our knowledge by utilizing Department of Transportation records, or perhaps setting up an experiment to count the number of vehicles passing over that bridge during a given period of time.

The Systems Engineer must identify which gaps need to be closed before moving forward with the system's design, weighing the amount of time and investment that is required to close the gaps. In some cases; modeling and simulation might be used, or prototyping or comparison to similar systems may also be adequate. In other cases, the time requirements and investment are too large to consider a closing of the gap, in which case the System Engineer may choose to accept the risk of a design decision in order to keep the program progressing.

Defining the Technical Solution

Upon completion of the System's Architecture, Requirements List and Verification Matrix, and once these Systems Engineer believes that the appropriate gaps in knowledge have been sufficiently closed or clarified, documentation may be finalized before the Physical Architecture is defined.

At this point, all information must be consolidated, including the Work Products and Requirements List, with a review of these documents for coherence and lack of conflicting requirements. This phase often requires revision of the System Architecture or adjustment to the Requirements List. When this has been completed, the System Architecture may be transferred to the Component Engineers for translation to Physical Architecture and detailed design. This is accomplished in a handoff meeting, where the Component Engineers are introduced to the System Architecture and the Systems Engineering team clarifies all uncertainties and questions. In many cases, the members of both teams are identical, but the role played by specific individuals on the team may shift as the focus of the work progresses through the development cycle.

Defining the Physical Architecture

Once handoff has occurred, the Component Engineering team utilizes the System Architecture as a blueprint for the Physical Architecture definition. Paying close attention to the construction of each functional block on the Architecture diagrams, the Component Engineers establish physical designs to accomplish the capabilities outlined by the Systems Engineering group.

The Component Engineers continue to develop the project utilizing standard Engineering practices and procedures to complete the decomposition phase of the V diagram and begin the re-composition phase. Their output completes the detailed design documentation, as well as the initial verification of the components and subsystems produced.

This concludes the System Development portion of the Systems Engineering cycle. The remaining responsibility of the Systems Engineering group constitutes the verification and validation of the resulting system produced.

Verification and Validation

The Verification and Validation of a system is the second major development area for the Systems Engineer. The Verification and Validation process constitutes the right-hand side of the V diagram of the systems process, and represents further re-composition of the system including the verification that we have completed the system as intended, and in a manner that will satisfy the customer.

The Verification and Validation process is comprised of three related concepts:

Verification – The process of evaluating a system or component to determine whether the products of a given development phase satisfy the conditions imposed at its start. Did we build the product correctly?

Validation – The process of evaluating a system or component during or at the end of the development process to determine whether it satisfies the specified requirements. Does the product do what it should?

Certification – A written guarantee that a system or component complies with its specified requirements and is acceptable for operational use. Is an outside expert happy enough with the product to stake his/her reputation on it?

Verification and Validation are typically conducted by the Systems Engineer in order to produce an efficient and effective design result. A fast design process only helps if you get it right. If you get it wrong, you spend far more money fixing problems because you have progressed farther into the design phase before you address them.

We must put Verification and Validation into context to fully understand the impact on the success of a system's design. This will also allow us to better understand the difficulty of accomplishing Verification and Validation on complex systems. First and foremost, we must understand:

No computer code / physical design can be proved to have no errors.
No documentation package can be proved to have no errors.

Verification and Validation is typically carried out not to assure that the system is perfect and in no need of future attention, but rather, that we are reasonably confident it will operate in the manner we have planned for a specified period of time.

This is often done throughout the development cycle, and perhaps without a great deal of thought. For example, when we are seeking to create a system that will improve upon an existing system, we may utilize a model of that system in order to simplify our approach.

If you are designing a heatsink, you may rely on thermodynamic equations and fin theory to begin your system design. However, you may not give a great deal of thought to the Validation of the equations with regard to real-world behaviors. In reality, these equations are not particularly predictive of the physical behavior of most heatsinks. Without understanding that we are relying on a numerical representation of reality in our modeling of the system, we may be missing an important element of the Validation of the system. Therefore we must validate the translation from the physical world to the numerical description before we begin any development at all. If we assume our model is correct and it is not, we will likely not produce a system of value to the customer.

Once the model of the system is complete, it can be used to design the physical entity for the heatsink product. If we were to compare the physical product with what we intended to design (based upon our model), we would be verifying that our design reflects the behavior predicted by model. Ultimately, we will examine the heatsink in operation, which would be the Validation that the design is effective in the operational context.

So in very simple terms we can clarify the difference between Verification and Validation:

Verification – Did we do what we said we would do?
Validation – Did we do what the customer needed?

Typically Verification and Validation is conducted at a level commensurate with the criticality of the system designed. For example, if our system affects critical performance that may result in a loss of life, a loss of prestige, or a financial or social loss, we would likely verify and validate at a higher integrity level then a system that did not demonstrate these characteristics. Many companies that do not produce mission-critical systems do a very low level of Verification and Validation. Companies that do have a more direct impact on safety or health may do a very rigorous Verification and Validation. An example of the former might be a software company that produces simple applications. An example of the latter might be an automotive company.

As discussed during the System Development phase, we have defined a large number of system requirements that are constructed in a specifically verifiable form. These were eventually entered into our RVTM, which creates for us a convenient mechanism by which to verify our system.

Verification is usually conducted throughout the System Development phase as each design phase is completed in the System Development. Typically Verification will take one of four potential forms, as shown in our traceability matrix. We will use one or more of these modes to check each of our requirements in order to complete our verification process. The four modes include:
Inspection – Do the characteristics of the component match the specification?
Analysis – Does a model predict that the system will perform as required?
Test – The component is placed in a simulated or actual operational use to check performance.
Demonstration – The entire system is demonstrated for the customer to ensure satisfaction.

Verification and Validation (V&V) are, by definition, ongoing activities. There is likely no endpoint to Verification and Validation challenges. This implies that it is possible to continue Verification and Validation long beyond any practical return. Therefore, in order to manage the expense incurred in the V&V process, we must examine the impact of cost and time with regard to our confidence level. This is shown in Figure 80. Based upon the testing integrity required, we will define appropriate limits for our Verification and Validation processes.

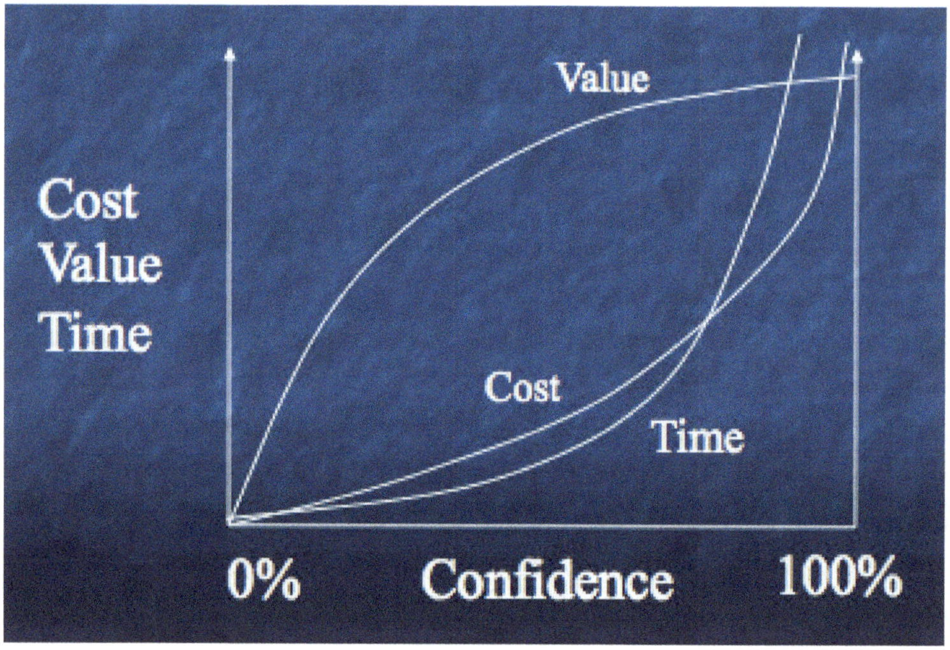

Figure 80: Confidence of V&V

Verification

In order to accomplish Verification, we need to ascertain whether the system has been created or built correctly. To begin, careful planning is required to get the most out of any testing and inspection process. Test planning is about defining standards for the testing process rather than describing the product tests.

The testing process will begin with unit testing, then module testing, then subsystem testing, then systems testing, and finally culminating in an acceptance test. This is representative of moving back up the right-hand side of the systems V diagram. The structure of a formal test plan should include:
The testing process
Requirements traceability
Tested items
Testing schedule
Test recording procedures
Hardware and software requirements
Constraints

Inspections and Testing are complementary and not contradictory Verification techniques. Inspections will check conformance of the system with the specification, but not conformance of the system with the customer's real requirements. Additionally, inspections cannot check non-functional characteristics; such as performance, usability, reliability and others.

In order to speed up the inspection or Verification of a system, check lists of common errors are typically used to help focus the inspection. In addition, experienced professionals are also used, relying upon their experience and expertise to judge the systems have been completed correctly.

Testing can take many different forms. For example, many organizations conduct hardware testing; essentially shake and bake testing of the system both statically and in operation. Some may also conduct stress and life testing, running a highly stressed system to find points of weakness or failure. Margin testing is also utilized, which overdrives a system in order to find the margin of safety. System-level testing focuses on usability tests and checks that the system can, in fact, be operated in the manner intended within reasonable time limits.

White–box testing examines all elements of a component or subsystem to ensure that each of the internal capabilities performs as expected. Black–box testing tests each of the functional blocks within the System Architecture to ensure that the system operates correctly. Typically, the process begins with white-box testing of all components, and progresses through black-box testing of consolidated components until the fully assembled system is both Verified and Validated. Ultimately, the entire system is tested as a black-box solution for the customer.

Validation focuses on the customer's acceptance of the product. In general, this is accomplished by demonstration of the system in operation. This type of testing is largely focused on demonstrating each of the established user needs from the outset of the System Development process.

As previously stated, V&V is a big can of worms. Since we cannot prove that the system, specification or requirements are complete or correct, we must instead come to some conclusion with regard to what confidence level we feel is appropriate prior to shipping the product. The general state of knowledge is that we will keep testing until it takes a long time to find the next defect. The definition of long time is dependent upon the organization.

Ultra dependable systems are becoming more commonplace. These are systems that we expect will never fail. For example, banking systems, healthcare systems, and transport systems fall into this classification. In general, a very rigorous V&V approach will be applied, and the testing will be conducted ten times longer than the acceptable mean time between failures (MTBF). This will likely ensure that our system will demonstrate ultra-dependable behavior.

For the assistance of the reader, a sample Verification and Validation template has been included in Appendix A.

Chapter 6: Resource Planning

As a leader in an Engineering department, it is as important to plan and schedule project work as it is to understand the appropriate staffing levels necessary to accomplish the tasks at hand. This implies that the Engineering leader must make a parallel effort with regard to resource scheduling and resource planning.

In the case of resource scheduling, a project plan is typically written with an eye towards understanding what resources and capabilities are available within the department to accomplish specific jobs and tasks. In general, these schedules are defined as though the project were being executed in a vacuum. By virtue of this, when multiple schedules are merged and combined, resources are often both dramatically over utilized in the case of talented employees and dramatically underutilized in the case of less talented employees. In some cases this will be more directly related to the shifting demands of skill areas within the organization as opposed to specific talents of employees.

It is recommended that the projects be scheduled and resource-loaded in the recommended project management manner to ensure the most efficient project plans are defined for execution. It will, however, be important for the Engineering leader to understand where the conflicts and utilization challenges occur within the department. Therefore, a secondary planning option is recommended to predict and plan for appropriate staffing levels within the organization.

A simple resource planning spreadsheet can be created to accomplish these tasks efficiently and effectively. To begin, we define all of the variables of interest to us related to the description of the specific assignments provided to the resources. This classification of variables as they relate to work assignments will be used to allow cross-referencing of the data in our resource planning process. As an example, you may choose to include the individual contributor's name, position, functional group, department, geographic area, project, and program on your list (these become your columns). The list is then created, describing every single work assignment provided within the Engineering organization. In most cases, this will constitute three or four work assignments per employee, except for those employees completely dedicated to specific efforts. An example of this is shown in Figure 81.

Staffing Plan

Name	Position Role	Program	May-15	Jun-15	Jul-15	Aug-15	Sep-15	Oct-15	Nov-15	Dec-15	Jan-16	Feb-16	Mar-16	Apr-16
			%FTE status for each month											
Abrahamson, Joel	Systems Engineer	S/L Product Sys Int	0.2	0.2	0.2	0.2	0.2	0.2	0.2	0.2	0.2	0.2	0.2	0.2
Abrahamson, Joel	Systems Engineer	Dent Reduction	0.1	0.1	0.1									
Abrahamson, Joel	Systems Engineer	Sharp	0.3	0.3	0.2	0.2	0.2	0.2	0.2	0.2	0.2	0.2	0.2	0.2
Abrahamson, Joel	Systems Engineer	Apple		0.1	0.2	0.2	0.2	0.2	0.2	0.2	0.2	0.2	0.2	0.2
Abrahamson, Joel	Systems Engineer	Invisibility Project	0.25	0.25	0.1	0.1	0.1	0.1	0.1	0.1	0.1	0.1	0.1	0.1
Abrahamson, Joel	Systems Engineer	Green/Black Belt	0.1	0.1	0.1	0.1	0.1	0.1						
Abrahamson, Joel	Systems Engineer	Field Support	0.2	0.1	0.1	0.1	0.1	0.1	0.2	0.2	0.2	0.2	0.2	0.2
Bell, Erin	Research Scientist	Photo Etch Development	1	1	1	1	1	1	1	1	1	1	1	1
Brearey, Robert	Systems Engineer	70 S/M 5 - 1 Validation	0.1	0.1	0.1	0.1								
Brearey, Robert	Systems Engineer	60 S/M 5 - 2 Validation	0.1	0.1	0.1	0.1								
Brearey, Robert	Systems Engineer	60 S/M 5 - 3 Validation	0.1	0.2	0.2	0.1								
Brearey, Robert	Systems Engineer	S/M Product Sys Int	0.2	0.2	0.2	0.2	0.25	0.25	0.25	0.25	0.25	0.25	0.25	0.25
Brearey, Robert	Systems Engineer	B^2 Sensor Developmen	0.2	0.2	0.1	0.1	0.1	0.1	0.1	0.1	0.1			
Brearey, Robert	Systems Engineer	Invisibility Project	0.2	0.2	0.2	0.1	0.1	0.1	0.1	0.1	0.1	0.1	0.1	0.1
Brearey, Robert	Systems Engineer	Field Support	0.1	0.1	0.1	0.2	0.4	0.4	0.4	0.4	0.4	0.4	0.4	0.4
Fried, Andrew	Systems Engineer	Invisibility Project	0.1	0.1	0.1	0.1	0.1	0.1	0.1	0.1	0.1	0.1	0.1	0.1
Fried, Andrew	Systems Engineer	CN Touch Testing	0.5	0.5	0.5	0.5	0.5	0.5	0.5	0.5	0.5	0.5	0.5	0.5
Fried, Andrew	Systems Engineer	Sharp	0.1	0.1	0.1	0.1	0.1	0.1	0.1	0.1	0.1	0.1	0.1	0.1
Fried, Andrew	Systems Engineer	Apple		0.1	0.1	0.1	0.1	0.1	0.1	0.1	0.1	0.1	0.1	0.1
Fried, Andrew	Systems Engineer	Percolation Modeling	0.1	0.1	0.1		0.1		0.1		0.1		0.1	
Fried, Andrew	Systems Engineer	90 Deg Pull Test	0.1	0.1										
Fried, Andrew	Systems Engineer	Field Support	0.1	0.1	0.1	0.1	0.1	0.1	0.1	0.1	0.1	0.1	0.1	0.1
Guo, Liping	Project Manager	TC Liner	0.25	0.2	0.2	0.2	0.2	0.2	0.2	0.2	0.2	0.2	0.2	0.2
Guo, Liping	Project Manager	HC liner	0.7	0.7	0.7	0.7	0.7	0.7	0.8	0.8	0.8	0.8	0.8	0.8
Guo, Liping	Project Manager	Green/Black Belt	0.2	0.2	0.2	0.2	0.2	0.2						
Kerfeld, Russell	Systems Engineer	60 S/M 5 - 3 Validation	0.2	0.1	0.1									
Kerfeld, Russell	Systems Engineer	B^2 Sensor Developmen	0.1	0.1	0.1									
Kerfeld, Russell	Systems Engineer	Invisibility Project	0.2	0.2	0.2	0.1	0.1	0.1	0.2	0.2	0.2	0.2	0.2	0.2
Kerfeld, Russell	Systems Engineer	Ag Migration	0.1	0.1	0.1	0.1	0.1	0.1	0.2	0.2	0.2	0.2	0.2	0.2
Kerfeld, Russell	Systems Engineer	Green/Black Belt	0.2	0.2	0.2	0.2	0.2	0.2						
Kerfeld, Russell	Systems Engineer	Field Support	0.2	0.2	0.2	0.2	0.2	0.2	0.4	0.4	0.4	0.4	0.4	0.4
Kerfeld, Russell	Systems Engineer	Touch Line Installation	0.1	0.1	0.1	0.1	0.1	0.2	0.2	0.2	0.2	0.2	0.2	0.2
Luhman, Wade	Systems Engineer	Photo Etch Development	0.8	0.8	0.8	0.8	0.8	0.8	0.8	0.8	0.8	0.8	0.8	0.8
Luhman, Wade	Systems Engineer	Field Support	0.2	0.2	0.2	0.2	0.2	0.2	0.2	0.2	0.2	0.2	0.2	0.2
Monson, Robert	Apps Eng Mgr	Administration	0.5	0.5	0.5	0.5	0.5	0.5	0.5	0.5	0.5	0.5	0.5	0.5
Monson, Robert	Apps Eng Mgr	Dent Reduction	0.05	0.05	0.05	0.05	0.05	0.05	0.05	0.05	0.05	0.05	0.05	0.05

Figure 81: Resource Planning Spreadsheet

Note that this spreadsheet includes the names of each employee, their workgroup, programs to which they contribute, and the percentage of time they will apply to each of these programs on a monthly basis. This is constructed in this manner in order to provide a twelve-month forward-looking description of expected work levels within the Engineering department.

Once the spreadsheet is completed, a separate page is created to define the staffing report. This is constructed as a pivot table and allows us to sort the information presented by the different variables available. In fact, this report is only limited by the number of variables you have included in columns on your staffing resource sheet. Having this pivot table allows for the examination, for example, of what programs each person is working on or how many total Mechanical Engineers are needed over the next twelve months. Each of these reports might be useful in a variety of ways, but they allow the Engineering leader to examine the staffing levels of the organization. In the event of conflicts and challenges identified, the leader can adjust the start and end points of projects to reduce those conflicts to manageable levels (shift projects to the left and right).

A few examples are included in the following diagrams. Note that the information can easily be sliced in a variety of ways in order to produce reports containing the information of interest to the planner.

Work Group	Applications Engineering					
		Data				
Name	Program	Sum of May-15	Sum of Jun-15	Sum of Jul-15	Sum of Aug-15	Sum of Sep-15
Abrahamson, Joel	Apple		0.1	0.2	0.2	0.2
	Dent Reduction	0.1	0.1	0.1		
	Field Support	0.2	0.1	0.1	0.1	0.1
	Green/Black Belt	0.1	0.1	0.1	0.1	0.1
	Invisibility Project	0.25	0.25	0.1	0.1	0.1
	S/L Product Sys Int	0.2	0.2	0.2	0.2	0.2
	Sharp	0.3	0.3	0.2	0.2	0.2
Abrahamson, Joel Total		1.15	1.15	1	0.9	0.9
Bell, Erin	Photo Etch Development	1	1	1	1	1
Bell, Erin Total		1	1	1	1	1
Brearey, Robert	60 S/M 5 - 2 Validation	0.1	0.1	0.1	0.1	
	60 S/M 5 - 3 Validation	0.1	0.2	0.2	0.1	
	70 S/M 5 - 1 Validation	0.1	0.1	0.1	0.1	
	B^2 Sensor Development	0.2	0.2	0.1	0.1	0.1
	Field Support	0.1	0.1	0.1	0.2	0.4
	Invisibility Project	0.2	0.2	0.2	0.1	0.1
	S/M Product Sys Int	0.2	0.2	0.2	0.2	0.25
Brearey, Robert Total		1	1.1	1	0.9	0.85
Brezinka, Rick	60 S/M 5 - 3 Validation	0.2	0.2	0.2	0.2	0.2
	Ag Migration	0.2	0.2	0.2	0.2	0.2
	B^2 Sensor Development	0.2	0.2	0.2	0.2	0.2
	CN Touch Testing	0.2	0.2	0.2	0.2	0.2
	Invisibility Project	0.2	0.2	0.2	0.2	0.2
	Touch Line Installation	0.1	0.1	0.1	0.1	0.1
Brezinka, Rick Total		1.1	1.1	1.1	1.1	1.1
Fried, Andrew	90 Deg Pull Test	0.1	0.1			
	Apple		0.1	0.1	0.1	0.1
	CN Touch Testing	0.5	0.5	0.5	0.5	0.5
	Field Support	0.1	0.1	0.1	0.1	0.1
	Invisibility Project	0.1	0.1	0.1	0.1	0.1
	Percolation Modeling	0.1	0.1	0.1		0.1
	Sharp	0.1	0.1	0.1	0.1	0.1
Fried, Andrew Total		1	1.1	1	0.9	1
Guo, Liping	Green/Black Belt	0.2	0.2	0.2	0.2	0.2
	HC liner	0.7	0.7	0.7	0.7	0.7
	TC Liner	0.25	0.2	0.2	0.2	0.2
Guo, Liping Total		1.15	1.1	1.1	1.1	1.1

Figure 82: Associate Assignments by Program

Work Group	Applications Engineering					
	Data					
Program	Sum of May-15	Sum of Jun-15	Sum of Jul-15	Sum of Aug-15	Sum of Sep-15	Sum of Oct-15
60 S/M 5 - 2 Validation	0.1	0.1	0.1	0.1		
60 S/M 5 - 3 Validation	0.5	0.5	0.5	0.3	0.2	0.2
70 S/M 5 - 1 Validation	0.1	0.1	0.1	0.1		
90 Deg Pull Test	0.15	0.15	0.05			
Administration	0.5	0.5	0.5	0.5	0.5	0.5
Ag Migration	0.3	0.3	0.3	0.3	0.3	0.3
Apple		0.2	0.3	0.3	0.3	0.3
B^2 Sensor Development	0.55	0.55	0.45	0.3	0.3	0.3
CN Touch Testing	0.7	0.7	0.7	0.7	0.7	0.7
Dent Reduction	0.15	0.15	0.15	0.05	0.05	0.05
Dry View Projects	1	1	1	1	1	1
Field Support	1.1	1	1	1.1	1.3	1.3
Green/Black Belt	0.5	0.5	0.5	0.5	0.5	0.5
HC liner	0.75	0.75	0.75	0.75	0.75	0.75
Invisibility Project	1	1	0.85	0.6	0.6	0.6
Percolation Modeling	0.15	0.15	0.15		0.1	
Photo Etch Development	3.7	3.7	3.7	3.7	3.7	3.7
S/L Product Sys Int	0.2	0.2	0.2	0.2	0.2	0.2
S/M Product Sys Int	0.2	0.2	0.2	0.2	0.25	0.25
Sharp	0.4	0.4	0.3	0.3	0.3	0.3
TC Liner	0.25	0.2	0.2	0.2	0.2	0.2
Touch Line Installation	0.4	0.4	0.4	0.4	0.4	0.5
Grand Total	12.7	12.75	12.4	11.6	11.65	11.65

Figure 83: Resource Needs by Program

Work Group	Applications Engineering					
	Data					
Position Role	Sum of May-15	Sum of Jun-15	Sum of Jul-15	Sum of Aug-15	Sum of Sep-15	Sum of Oct-15
Applications Engineer	1	1	1	1	1	1
Apps Eng Mgr	1.2	1.2	1.2	1	1	1
Project Manager	1.15	1.1	1.1	1.1	1.1	1.1
Research Scientist	1	1	1	1	1	1
Research Technician	2.1	2.1	2.1	2.1	2.1	2.1
Systems Engineer	6.25	6.35	6	5.4	5.45	5.45
Grand Total	12.7	12.75	12.4	11.6	11.65	11.65

Figure 84: Resource Needs by Function

This method produces a highly flexible tool that allows great insight into the resource challenges faced by engineering organization. It is also used to justify staffing levels within the group. As an example, if the organization is required to accomplish those programs listed in Figure 83, then the staffing requirements for the Engineering department are fixed based upon this plan. If, however, the programs must be completed within a specific window, it is possible that we could shift programs right and left in order to maintain relatively stable staffing levels within the department.

Chapter 7: Engineering Titles and Career Paths

When managing an Engineering organization, it is important to recognize the value of the individuals who are members of the department. In general, technical professionals are motivated by a wide variety of offerings within the organization, and as a general rule many of the technical professionals are motivated by a desire to be successful and effective.

By virtue of this, the Engineering leader must plan for the career progression of his or her associates within the organization. This enables several things to happen. First, it provides a clear structure for relative position within the organization. Second, it delineates different performance parameters for each level within the structure, thereby allowing assessment of different Engineering levels within the organization. Third, it enables the construction of a salary structure that ensures individuals with greater responsibility have greater levels of reward. Fourth, it provides direct insight to the associates of the expectations of each progression forward within the title structure, thereby allowing them to plan their personal development.

Many different structures have been established to define these different levels within the organization, and in general it is preferable to limit the number of different levels to a practical value. One might ask what a practical value is for these levels. To establish that, one must consider where the average individual will finish out their careers.

For example, if the majority of your Engineering staff will retire at a level equivalent to a Project Manager, one must consider that a typical career will extend approximately forty years, and Engineering professionals will expect promotion early in their careers. Therefore, one level must be planned for the transition from an entry-level Engineer to an independently operating Engineer. This promotion will be expected to occur within the first three years of employment. If the Project Manager is at a level three, for example, this implies that the typical associate could expect only one more promotion prior to retirement. If the Engineering leader would prefer to provide for promotions on an approximately a ten-year basis, more levels would be appropriate.

A relatively common model followed by many large government contractors establishes five distinct levels within the Engineering organization. There are executive and corporate positions beyond this not delineated in this construction, but the vast majority of the Engineering staff will fall into this classification. An example of this classification is shown in Figure 85. Please note the title structure is separated by function as well, allowing differentiation in salary bands for different disciplines if necessary.

Level	Mechanical	Electrical	Software	Systems	Manager	Designer	Technician
1	Design Engineer, Mechanical	Design Engineer, Electrical	Design Engineer, Software	Design Engineer, Systems		Designer	Technician
2	Senior Engineer, Mechanical	Senior Engineer, Electrical	Senior Engineer, Software	Senior Engineer, Systems		Senior Designer	Senior Technician
3	Staff Engineer, Mechanical	Staff Engineer, Electrical	Staff Engineer, Mechanical	Staff Engineer, Systems		Staff Designer	Staff Technician
4	Project Engineer	Project Engineer	Project Engineer	Project Engineer	Manager		
5	Principal /Chief Engineer	Principal / Chief Engineer	Principal Engineer / Architect	Principal Engineer / Architect	Director		

Figure 85: Engineering Title Structure

In addition to the Engineering titles, equivalent levels are defined for First-line Managers and Directors within the structure. Many Engineering organizations also employ both Designers and Technicians, and these will typically have fewer levels of progression. Should the need arise for a Designer or Technician to progress beyond level three (the top level), consideration would be given to transferring the associate to the Engineering career path.

Naturally, salary bands are associated with the different levels within an Engineering organization, allowing associates to understand how they will progress in title as well as financial compensation. A generic example of this is shown in Figure 86.

Level	Mechanical	Min	Mid	Max
1	Design Engineer, Mechanical	$48,000	$60,000	$72,000
2	Senior Engineer, Mechanical	$60,000	$75,000	$90,000
3	Staff Engineer, Mechanical	$68,000	$85,000	$102,000
4	Project Engineer	$80,000	$100,000	$120,000
5	Principal /Chief Engineer	$96,000	$120,000	$144,000

Figure 86: Salary Bands Relative to Level

With this construction, all members of the Engineering organization may understand their relative positions within the organization, and the question will naturally arise regarding what the expectations are for the next level in their career trajectories. To answer this question, it is equally important to define specific job expectations in the form of a job description for each level. A generic example of this is provided.

Engineer level 1 (Engineer)

Entry-level professional requiring Bachelor's Degree in Engineering and no experience, or the equivalent (to a degree) in appropriate education and experience. Works under close supervision; receives specific and detailed instructions for required tasks and results expected. Performs a variety of routine tasks, which provide experience and familiarity with Engineering staff, methods, practices, and programs. Usually assumes no responsibility for direction of others. Limited exercise of judgment required when less common methods or procedures are necessary. Assignments may include higher-level work for training/developmental purposes. Supervisor screens assignments for unusual complexities and selects non-routine techniques and procedures to be applied. Receives close supervision on new aspects of assignments.

Qualifications:

B.S. in Engineering or a related field preferred, other educational levels accepted with commensurate experience.

AutoCAD/Inventor preferred, comparable 3D Modeling experience required.

Demonstrates ability to design machine structures in accordance with accepted design practices.

Preferred skills include electrical / mechanical drive sizing and hydraulic and pneumatic circuit design.

Typical Duties:

Develops and reviews drawings, specifications or proposals to insure the quality of products.

Participates in new product design teams using stage-gate processes.

Recommends correction actions in customer problem resolution as required.

Supports other functional areas as requested, including Manufacturing, Sales, Service, HR, Finance, Installation and Training.

Maintains accurate and timely product information.

May travel to customer site to assist with machine start-up, installation, trouble-shooting and customer training as required.

Provides complete drawing packages and documentation to Product Managers and Manufacturing, as required.

Researches and develops new technologies to enhance production capabilities and produce product and process improvements.

Performs other work-related duties as assigned.

Engineer level 2 (Senior Engineer)

Independently evaluates, selects, and applies standard engineering techniques and procedures while using judgment when making minor adaptations and modifications. Assignments have clear and specific objectives and require investigation of limited number of variables. Receives instructions on specific assignment objectives, complex features, and possible solutions. Assistance given for unusual problems and normally reviewed for application of sound professional judgment. Performs work involving conventional plans, investigations, surveys, structures, or equipment with relatively few complex features for which there are few precedents

Qualifications: B.S. in Engineering or a related field preferred, other educational levels accepted with commensurate experience, typically 5-7 years applicable work experience.

AutoCAD/Inventor preferred, comparable 3D Modeling experience required.

Demonstrated ability to design machine structures in accordance with accepted design practices.

Preferred skills include electrical / mechanical drive sizing and hydraulic and pneumatic circuit design.

Demonstrated ability to lead a group of cross-functional Engineers and Designers to the successful completion of a design and manufacture of mechanical machine, including mechanisms, hydraulics and pneumatic components and assemblies.

Typical Duties:

Develops and reviews drawings, specifications or proposals to insure the quality of products.

Participates in new product design teams using stage-gate processes.

Recommends correction actions in customer problem resolution as required.

Supports other functional areas as requested, including Manufacturing, Sales, Service, HR, Finance, Installation and Training.

Maintains accurate and timely product information.

May travel to customer site to assist with machine start-up, installation, trouble-shooting and customer training as required.

Provides complete drawing packages and documentation to Product Managers and Manufacturing, as required.

Researches and develops new technologies to enhance production capabilities and produce product and process improvements.

Performs other work-related duties as assigned.

Engineer level 3 (Staff Engineer)

Fully competent engineer in all conventional aspects of subject matter or functional area of assignments; plans and conducts work requiring judgment in independent evaluation, selection, and substantial adaptation/modification of standard techniques, procedures, and criteria. Devises new solutions to problems encountered. Independently performs most assignments with instruction only regarding general expected results. Receives technical guidance for unusual or complex problems and supervisory approval of proposed project plans. May supervise a few engineers and/or technicians on a project basis.

Qualifications: B.S. in Engineering or a related field preferred, other educational levels accepted with commensurate experience, typically >10 years applicable work experience or an M.S. degree with 3-5 years applicable experience.

AutoCAD/Inventor preferred, comparable 3D Modeling experience required.

Demonstrated ability to design machine structures in accordance with accepted design practices.

Preferred skills include electrical / mechanical drive sizing and hydraulic and pneumatic circuit design.

Typical Duties:

Develops and reviews drawings, specifications or proposals to insure the quality of products.

Participates in new product design teams using stage-gate processes.

Recommends correction actions in customer problem resolution as required.

Supports other functional areas as requested, including Manufacturing, Sales, Service, HR, Finance, Installation and Training.

Maintains accurate and timely product information.

May travel to customer site to assist with machine start-up, installation, trouble-shooting and customer training as required.

Provides complete drawing packages and documentation to Product Managers and Manufacturing, as required.

Researches and develops new technologies to enhance production capabilities and produce product and process improvements.

Supervises and leads Engineers, as they provide technical support for the design, manufacture, sales, installation, and service of all Park Industries products and services.

Performs other work-related duties as assigned.

Manager (Project Manager / Functional Manager)

Has full responsibility for interpreting, organizing, executing and coordinating assignments. Plans and develops engineering projects concerned with unique or controversial complexities which have important impact on major company programs. This involves exploration of subject area, definition of scope, selection of areas for investigation, and development of novel concepts. Acts as technical liaison to individuals within and outside the organization with responsibility to act independently regarding technical matters pertaining to individual's field. Supervision received is essentially administrative, with assignments given in broad terms concerning general objectives and limitations. Typical duties and responsibilities include one or more of the following: 1) plans, organizes, and supervises work of staff of engineers and technicians; 2) as individual researcher, consultant, or staff specialist, conceives plans and conducts research in areas of considerable scope and complexity.

Makes authoritative decisions and recommendations having significant impact on engineering challenges and related areas within the company. Negotiates critical and controversial issues with top-level engineers and officers of the company. Individuals at this level demonstrate a high degree of creativity, foresight, and mature judgment in planning, organizing, and guiding long-term engineering programs and activities of outstanding novelty and/or importance. Receives general administrative direction. Is responsible for one or more programs of such diversity and scope to be of critical importance to overall company objectives.

Provides overall supervision to individuals to assure that technical, administrative, man-hour, and schedule targets of the group are met within the framework of established corporate or organizational policy and in accordance with applicable professional standards, design-control procedures, and corporate or organizational procedures and guidelines. Interfaces with other managers to assure staffing is adequate for planned work, interfaces with Project and Program Managers to assure work is completed as planned. Accomplishes functional manager duties for group, including regular staff meetings and administrative tasks.

Qualifications: B.S. in Engineering or a related field preferred, other educational levels accepted with commensurate experience, typically >10 years applicable work experience or an M.S. degree with 3-5 years applicable experience.

Demonstrated ability to design machine structures in accordance with accepted design practices.

Preferred skills include electrical / mechanical drive design and hydraulic and pneumatic circuit design.

Leadership experience required.

Basic understanding of AutoCAD/Inventor/3D Modeling required.

Familiarity with products preferred.

Typical Duties:

Develops and/or reviews drawings, specifications or proposals to insure the quality of products.

Participates in new product design teams using stage-gate processes.

Recommends correction actions in customer problem resolution as required.

Supports other functional areas as requested, including Manufacturing, Sales, Service, HR, Finance, Installation and Training.

Maintains accurate and timely product information.

May travel to customer site to assist with machine start-up, installation, trouble-shooting and customer training as required.

Provides complete drawing packages and documentation to Product Managers and Manufacturing, as required.

Researches and develops new technologies to enhance production capabilities and produce product and process improvements.

Supervises and leads Engineers, as they provide technical support for the design, manufacture, sales, installation, and service of all park industries products and services.

Assures that all product documentation is accurate and complete.

Assures that all product changes are processed thru the ECO system.

Manages Engineering Support team so as to allow Product Support engineers (Subject matter experts), to be assigned to New Product Development projects as required, and assures that they are cross-trained to allow team members to be assigned to the New product development projects as required.

Is responsible for the functional management of the support personnel reporting, which includes personnel selection, training, mentoring and management of work assignments, personnel appraisal and the resolution of issues within the group.

Is responsible for the technical accuracy of the work within his or her team.

Establishes administrative policies, procedures, and standards as required.

Develops new products focusing on ability to manufacture, to deliver product function and product safety.

Provides project management expertise as required, and assures all long term projects are appropriately planned and executed.

Performs other work-related duties as assigned.

Director (Director / Principal Engineer)

Makes authoritative decisions and recommendations having significant impact on extensive engineering and related activities of company. Negotiates critical and controversial issues with top-level engineers and officers of their own and other organizations and companies. Individuals at this level demonstrate a high degree of creativity, foresight, and mature judgment in planning, organizing, and guiding extensive engineering programs and activities of outstanding novelty and/or importance. Receives general administrative direction. Is responsible for multiple programs of such diversity and scope to be of critical importance to overall company objectives.

Provides overall supervision to Department to assure that technical, administrative, man-hour, and schedule targets of Department are met within framework of established corporate or organizational policy and in accordance with applicable professional standards, design-control procedures, and corporate or organizational procedures and guidelines. Responsible for long-term staffing requirements and technology forecasting. Accomplishes functional manager duties for group, including regular staff meetings and administrative tasks.

Qualifications: B.S. in Engineering or a related field preferred, other educational levels accepted with commensurate experience, typically >20 years applicable work experience or an M.S. degree with >7 years applicable experience or PhD degree.

Demonstrated ability to design machine structures in accordance with accepted design practices.

Preferred skills include electrical / mechanical drive sizing and hydraulic and pneumatic circuit design.

Demonstrated ability to design machine structures in accordance with accepted design practices.

Preferred skills include electrical / mechanical drive design and hydraulic and pneumatic circuit design.

Leadership experience required.

Basic understanding of AutoCAD/Inventor/3D Modeling required.

Familiarity with products preferred.

Typical Duties:

Develops and/or reviews drawings, specifications or proposals to insure the quality of products.

Leads and participates in new product design teams using stage-gate processes.

Supports other functional areas as requested, including Manufacturing, Sales, Service, HR, Finance, Installation and Training.

Responsible for maintaining accurate and timely product information.

Will travel to customer site to assist with machine start-up, installation, trouble-shooting and customer training as required.

Provides completed drawing packages and documentation to Product Managers and Manufacturing, as required.

Researches and develops new technologies to enhance production capabilities and produce product and process improvements.

Supervises and leads Engineers, as they provide technical support for the design, manufacture, sales, installation, and service of all products and services.

Manages Engineering Support team so as to allow Product Support engineers (Subject matter experts), to be assigned to New Product Development projects as required, and assures that they are cross-trained to allow team members to be assigned to the New product development projects as required.

Is responsible for the functional management of the support personnel reporting, which includes personnel selection, training, mentoring and management of work assignments, personnel appraisal and the resolution of issues within the group.

Is responsible for the technical accuracy of the work within his or her team.

Establishes administrative policies, procedures, and standards as required.

Develops new products focusing on ability to manufacture, to deliver product function and product safety.

Provides project management expertise as required, and assures all long term projects are appropriately planned and executed, and trains other associates as needed.

Performs other work-related duties as assigned.

Chapter 8: Organizational Design

While it is generally true that any organization will function effectively as long as its members want it to, there are some benefits to reflecting on an appropriate organizational construct for the type of work that is to be performed. Some organizational designs will be better than others at facilitating effectiveness and efficiency within the organization.

It is therefore appropriate to consider first a taxonomy describing an organization in design (or behavioral) terms prior to considering the most commonly accepted organizational structures.

Many differing taxonomies exist to define organizational structure and job fit. Robbins defines six key aspects in his taxonomy:

Work Specialization

This relates to the division of labor. Typically efficiencies can be achieved by higher work specialization.

However, there can be limited usefulness, because with a high degree work specialization we typically find:

Lower job satisfaction

Higher boredom and fatigue

Therefore, we must consider both the advantages and disadvantages in deciding if this is a critical aspect in our organization.

Departmentalization

This relates to how we choose to separate our organizations. This may include:

Functional divisions

Product divisions

Geographic divisions

Process departments

Customer aligned departments

We must be careful when we create divisions within the organization, as this will virtually guarantee conflicts across the boundaries we create.

Chain of Command

This is the amount of authority bestowed by the organization upon an individual, and includes the unity of command principle. If we have a strong chain of command, we have an unbroken line of authority from the top to the bottom of the organization. This is considered good from an ambiguity standpoint, but not so good from the autonomy standpoint. If we have a weak chain of command, individuals within the organization may have multiple persons providing them with direction on a daily basis. This is commonly referred to as the 'two-boss problem', largely due to the fact that these work directions will often contain conflicting information. This must be carefully managed within a weak chain of command type organization, or work efficiency will suffer.

Span of Control

How many subordinates may be effectively managed? Should we utilize a span of four, or a span of eight? Either of these will result in dramatically different organizational structures, and require dramatically different numbers of people to accomplish our managerial duties.

It has been suggested that three to seven direct reports is typically ideal. However, we must consider the level of autonomy of the individuals before assessing an appropriate span of control. In cases where the associates are capable of a high amount of autonomy and independence, more direct reports can be managed. In cases where those direct reports need more managerial attention, we may choose to use a smaller span of control. It is somewhat common in Engineering departments to see a span of control ranging from three to forty direct reports, and arguments can be made supporting the use of any design in between.

Centralization

This relates to the degree to which decisions are made at lower levels of the organization. Higher decentralization means decisions are made in the trenches, which is considered good because it is done closer to the subject matter expertise. This will result in a better handle on the context of the decision and will be made by those to whom it has the greatest impact – this is empowerment. However, it can create a more interdependent structure, which will result in a higher degree of Chaos within the organization. Without clear strategic goals, this can also result in the production of a muddled effort.

Formalization

This relates to the degree of standardization in job duties. High formality means that job duties are highly standardized, whereas low formality results in greater autonomy for the employee. These considerations will result in higher or lower levels of Chaos within the organization as well.

When considering the appropriate structure for an Engineering organization, it is useful to consider this taxonomy and establish which of these factors are of importance and which are not. As an example, in an organization with a highly dynamic work environment requiring rapid decision-making, we would likely select a decentralized approach in order to support the type of organizational behavior we desire. As a further example, if we have relatively inexperienced managers, we may choose to utilize a smaller span of control until they become better acquainted with their duties.

Once we have established these crucial factors to consider in our organizational design, we can reflect upon common organizational structures, assessing whether they fit our needs for the desired behaviors within the organization. We will examine several of these structures for consideration. Most large organizations are made up several of these structures simultaneously.

Authority

The power granted to individuals to direct resources in the organization

Responsibility

Obligation of individuals to complete assignments and tasks

Accountability

Being totally answerable for satisfactory completion of an assignment

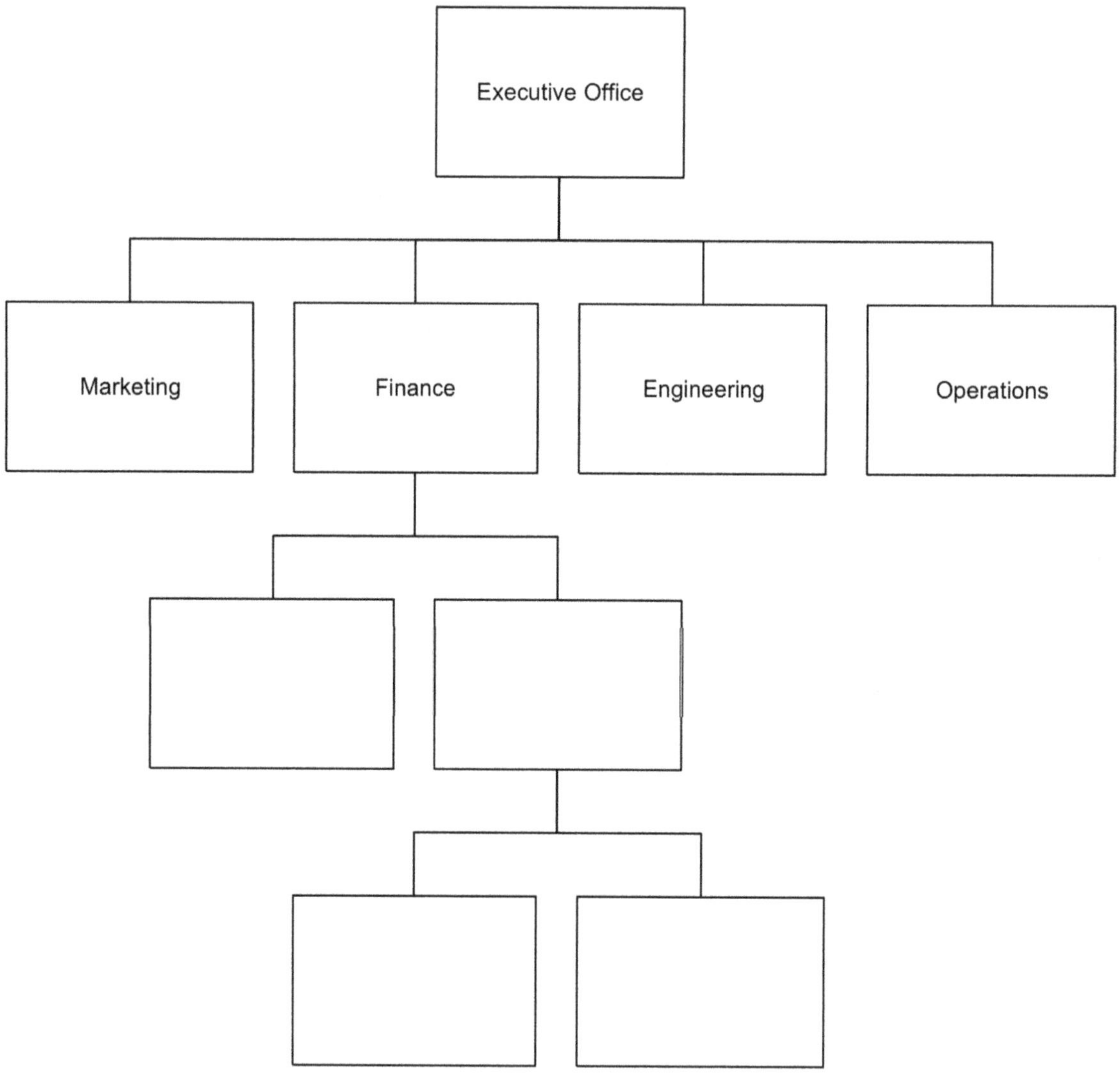

Figure 87: Traditional Structure

The traditional structure is one in which there is a very strong chain of command, where every associate in the organization has a single boss. This provides a very concrete assignment and task structure, and originally was created in military organizations for this reason.

The advantages of the traditional structure include easier budget and cost control structures, as the efforts do not span many boundaries. Continuity in the functional discipline often will create significant Engineering standards or policies that can drive consistency within the development process. Last, communications are driven largely in-line with the reporting structure, so all individuals are invited to a single staff meeting rather than to multiple meetings to hear similar information.

and Manufacturing, Sales and Manufacturing, Marketing and Engineering, and many others. The liaison is tasked with the responsibility to make sure that their assigned departments work effectively together.

While this construction is a major improvement with regard to conflict resolution within the organization, integration and coordination are still problematic. A strong liaison personality will often serve to resolve these differences with very little wasted time and paperwork, but it cannot be guaranteed that this will be the case.

The major disadvantage with this organizational design is that we have hired an entire department of personnel to accomplish what we hired all of the other personnel to accomplish in the first place. This duplication of effort is seen as wasteful and unnecessary in an organization that should be able to work effectively across functional boundaries.

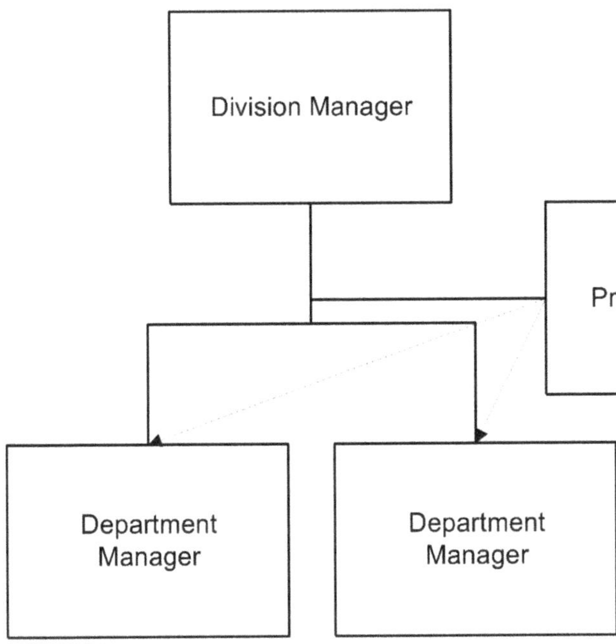

Figure 90: Line-Staff Structure

The Line – Staff organization was designed with the intention of having a Project Manager to oversee all details of a given project. This was good news from the Project Management perspective; unfortunately, when these individuals were positioned within the organization they were placed above the Department Managers.

From the perspective of the Project Manager, this enabled direct authority over tasking within the different divisions in the organization. From the perspective of the Department or Line

Manager, this resulted in the addition of many new bosses. As might be expected, if there was only a single Project Manager within the organization this might work effectively. However, when one considers that we might have as many as fifty Project Managers within the organization the ensuing work direction provided to the Department Managers becomes completely untenable. Therefore, in former cases we had the Project Manager positioned too low within the organization and in this case we have positioned them too high within the organization.

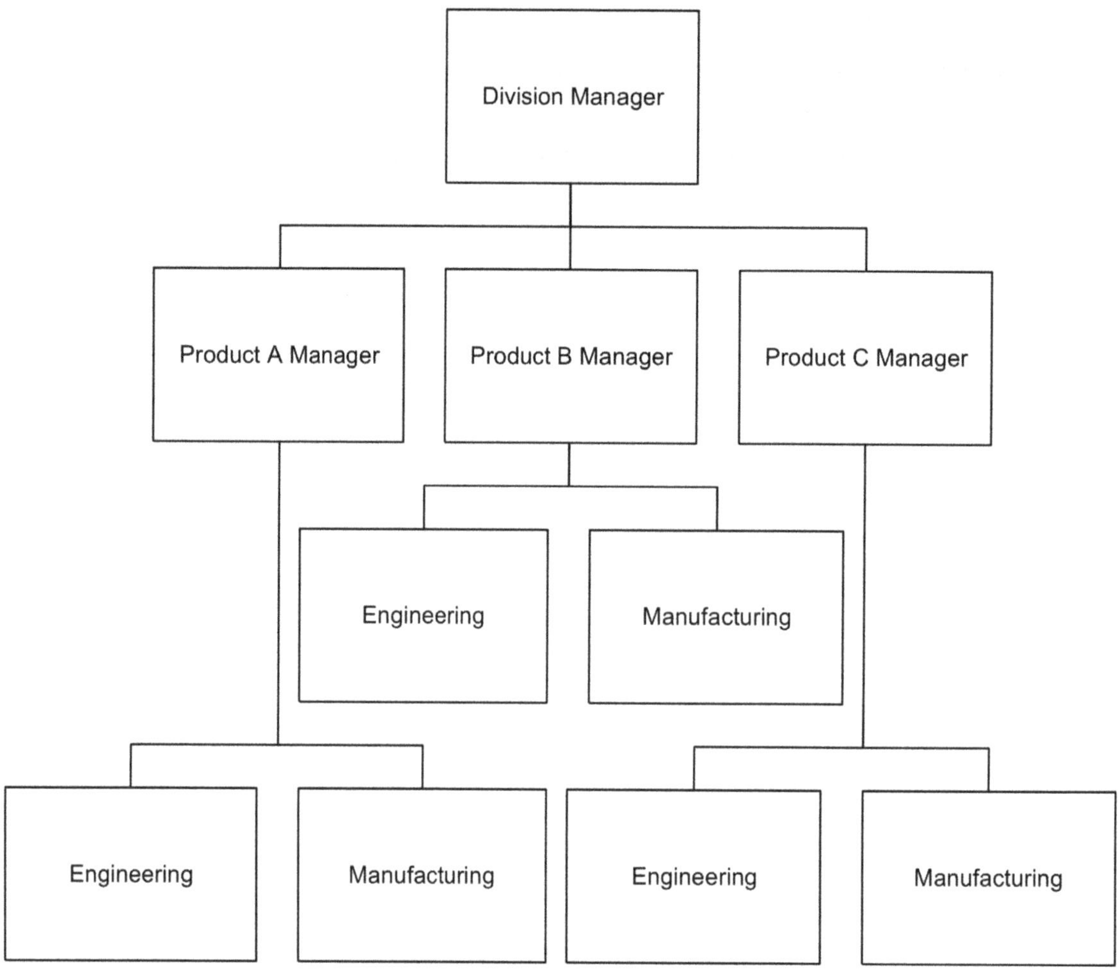

Figure 91: Product Structure (Brand Management)

In a product-structured organization, we find an ideal world for project management. In this particular case, if you have a project within the organization you also are assigned all functional resources necessary to complete that project. Therefore all functions direct report to the Project Manager.

Disadvantages of this system are largely associated with dramatically increased overhead costs and an inability to share resources across functional boundaries. In general, this organizational structure is equivalent to brand management as practiced by organizations such as General Mills. While this is extremely effective for General Mills, it can have considerably less impact on smaller organizations without adequate resources to support such an overbearing structure.

Other challenges associated with this structure include:
A tendency to keep people on the project after they are needed
Standards and internal technologies suffer from weaker functional groups
Top-level ordination amongst divisions is required
Lack of technical interchange between products
Lack of career opportunities for personnel

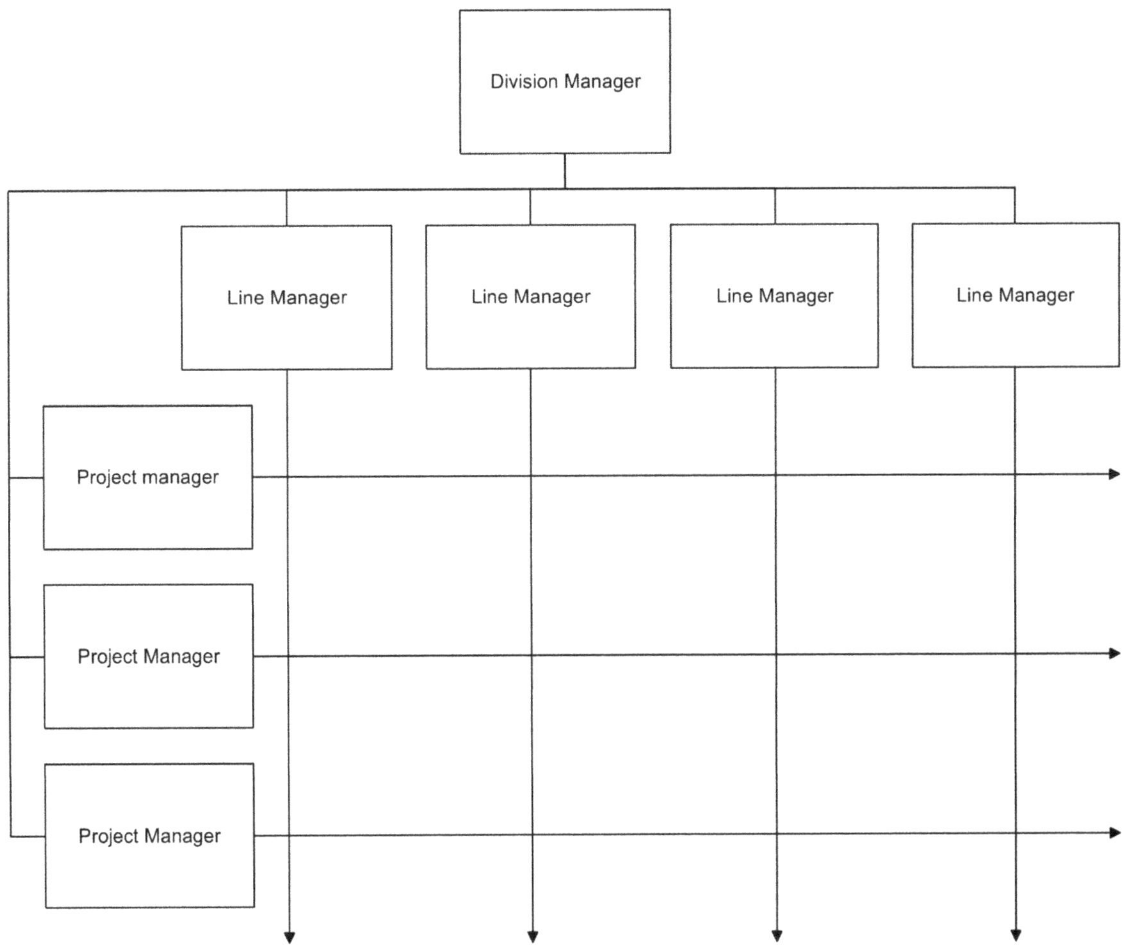

Figure 92: Matrix Structure

The matrix structure was developed in order to address the rapidly changing dynamic environments typical in most organizations. It is defined on three independent levels, including the Division, Functional (or Line), and Project Manager. Each of these responsibility areas carries with it different priorities and needs.

In an ideal situation, the matrix should operate effectively. The Division Manager considers the challenges to the organization, defining strategically what work needs to be accomplished in order to maintain efficiency and effectiveness for the organization. This strategy is disseminated to the Project Managers, who in turn develop projects and plans that would be appropriate implementations of the division's strategy. This is the tactical responsibility within the organization. Once the project plans are complete, the Project Managers (who are of the same authority level as the Line Managers) negotiate with the Functional Managers for resources. This is the task responsibility within the organization. Once the negotiation has taken place, the Functional Managers can communicate back to the division level regarding what staffing levels are required in order to complete the work as specified.

One challenge, however, that appears in this structure is related to the lack of formal definition regarding the responsibilities and duties of each of these positions. Due to this, many of the required duties get performed while some languish, as it is unclear where the responsibility resides.

A second challenge occurs when the Division Manager is unclear on the strategic focus of his or her position, and instead chooses to meddle in the operation of the matrix. This can often be seen in a fully loaded matrix organization that has been assigned several new 'hot' projects. The inclusion of these projects will often disrupt the precarious balance between the strategic, tactical, and task responsibilities within the matrix structure.

In highly complex organizations, there is often a need for a third dimension with regard to project developments, and this is accomplished through the use of Matrix Layering. This can be seen in Figure 93.

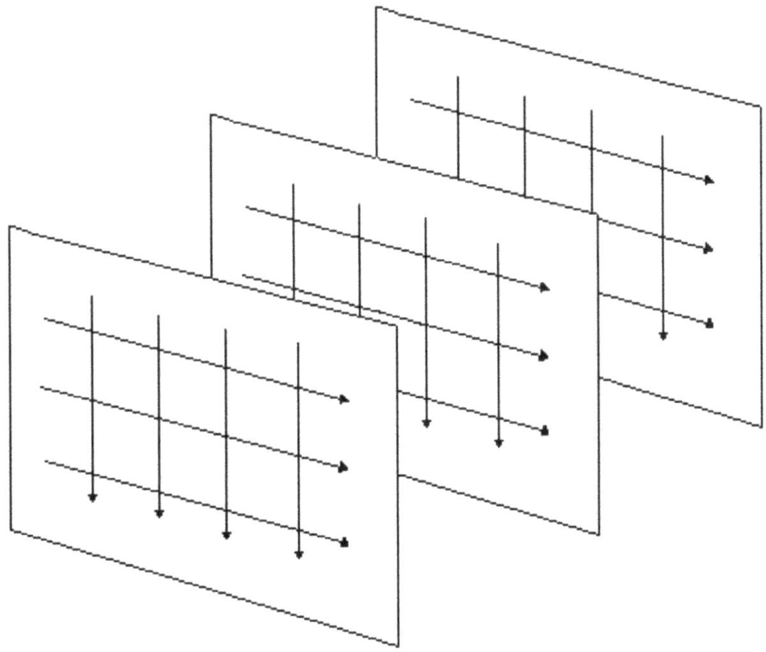

Figure 93: Matrix Layering

The layering of responsibility matrices is useful to assure that a high degree of flexibility occurs within the organization to size development teams appropriately for development efforts. As might be expected, this is also a highly volatile organization, without clear lines of responsibility. Due to this, a re-envisioning of this construction has been developed.

The cross-functional team is essentially a matrix organization where the semantics and naming conventions are designed to support the autonomous nature of the system. In the case of a matrix, it is very tempting for the Division Manager to meddle in the projects. In the case of the cross-functional team, the formation of the team precludes the ability to meddle in its performance. Therefore, by utilizing a matrix structure but by naming it differently, we change the perception of the organization and lend it a much higher degree of potential success. Cross-functional teams are now one of the most common organizational structures in use.

It is the responsibility of the Engineering leader to design an appropriate structure to accomplish the strategic, tactical and task oriented objectives of the organization.

Chapter 9: Performance Management and Appraisal Systems

Performance Appraisal is a crucial aspect of managing an Engineering organization. Regardless of the performance appraisal system presently in place within the organization, the Engineering leader must understand the importance of providing timely and meaningful feedback to the employee in order to produce a vibrant and effective workforce.

Why do we care about performance appraisal or understanding what behaviors are appropriate for individuals to succeed in the organization? There is a very simple rule that provides a great deal of insight into the behaviors we can expect from individuals within an organization:

You get what you reward!

Individuals are very adept at connecting behaviors and rewards within the organization. By virtue of this, the Engineering leadership must take great care to assure that their responses to associates behavior are aligned with intentions within the department. As an example, if the leadership believes coming in early is important, individuals who come in early must be rewarded. This encourages more of the same type of behavior. However, most of our organizations do a particularly poor job of aligning their reward systems with expected behaviors. If coming in late is discouraged, some punishment should be associated with the behavior. However, most organizations do not reward good behavior, nor do they punish bad behavior.

For example, imagine you have two Project Managers who are both fully planned. One always gets her work done efficiently, while the other never completes his work on time. When a hot new project comes in, the managers' leader must assign it to one of the Project Managers. The seemingly obvious choice would be the good Project Manager; however, this essentially punishes her for doing a good job. Now she is over-allocated and needs to spend extra hours at work. The poor manager is rewarded, and doesn't have to spend extra time at work. What behaviors can we expect to see in the future? The poor manager is perfectly satisfied with events; the good manager has probably realized that in order to spend more time with her family, she must do a poorer job at work. This situation is exactly the opposite of what we desire in our employees.

There are many examples in organizations where the appropriate behavior is rewarded, and many examples were inappropriate behavior is rewarded. We must think through what is appropriate within our organizations as an acceptable behavior for a typical employee. Once we have established an appropriate set of behaviors we must establish appropriate metrics and rewards that will correlate well with the desired performance. This should result in the appropriate motivation for appropriate behaviors.

Performance Appraisal is necessary to:

1. Make decisions about personnel
2. Define training and development needs
3. Assess training and development
4. Give the employees feedback on their performance
5. Define reward allocations

The basis of linking motivations and behaviors is grounded in behavioral theory, which defines the link between performance and reward. Several theories exist related to this concept, including Goal-Setting Theory, Reinforcement Theory, Equity Theory, Attribution Theory, and Expectancy Theory. One or more of these theories may be used to understand the decisions made by individuals within the organization regarding their behavior. With proper alignment of rewards and behaviors, we can coach associates to learn how to choose desired behaviors within the organization when conducting their daily tasks.

To make this clear to the individual, we will include it as the approach used for performance appraisal on an ongoing basis. Therefore, we must consider several domains in our process approach:

1. What to evaluate
2. Who should evaluate
3. How to evaluate
4. How to reward

We will consider each of these in turn.

What to evaluate

This element is perhaps the most difficult of all the issues that we must deal with in assessing people's behavior. We must give stringent thought to the types of behaviors that will be appropriate for us to be assured of success within the organization. So, as an example, we can consider a job description, and spend a few moments reflecting upon the appropriate behaviors for the job's requirements. For this example, we will consider the position of a musician within an orchestra. If we were to make a quick list of the measures we will use for performance appraisal, they might look like this:

Maintains professional skill level with instrument
Performs volunteer activities with orchestra
Is recognized within the orchestra as a ticket draw
Maintains good relationships with other members of the orchestra and director
Acts as a spokesperson for the orchestra
Shows up prepared for rehearsal
Is always on time and ready to perform

We find that it is actually quite challenging to create a list of metrics that will be useful in conducting a performance appraisal. In fact, we must consider how associates might accomplish the metrics set out for appraisal without performing at a high level within the organization. In other words, how could an associate manipulate these results in order to generate a better appraisal then might be deserved? With appropriate thoughts and metric design, we should be able to minimize this impact on our appraisal system.

Once we have completed this task we can assess the quality of the metrics we have defined. Typically we find that these metrics fall within three general classes:

Task Outcomes - the ends
Behaviors - number of times late, number of hours worked...
Traits - attitudes, knowledge, skills

What is appropriate is to define those behaviors we will value in the organization, because they directly contribute to better job performance and, in the long run, to organizational success.

However, we also must consider whether we have explored all of the necessary behaviors that we seek within the organization. To accomplish this, we will make use of an accounting tool that should help to focus our efforts. This is known as the Personal Scorecard. An example of this can be seen in Figure 94.

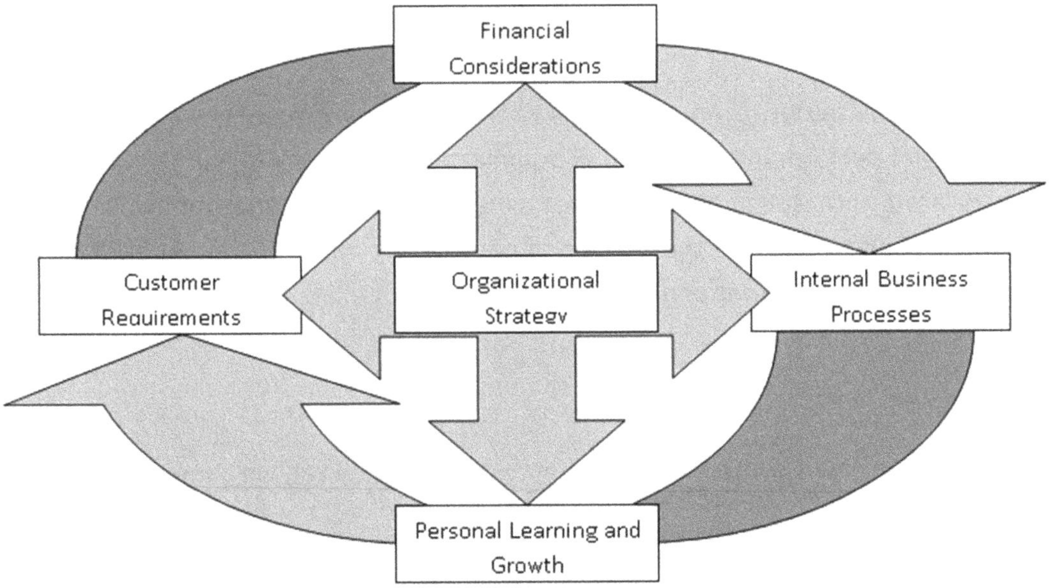

Figure 94: The Personal Scorecard

The Personal Scorecard is based upon The Balanced Scorecard, which is commonly used in managerial accounting systems. When we examine the list of performance metrics we previously generated, we noticed that we did not complete a cross-section of behaviors that might be desired by the organization, and instead generally focused in one or two areas such as bottom line or interpersonal skills. If we examine the Personal Scorecard, we notice that we must first define our organizational strategy, and construct metrics that will be in accordance with that strategy. Secondly, we need to make sure that we have several metrics in each of the four areas of interest on the scorecard. These include:

Financial considerations – how does this position impact the bottom line?
Customer requirements – how does our customer perceive our organization?
Personal learning and growth – what is this individual doing to improve himself or herself?
Internal business processes – what is this individual doing to improve our organization?

We should therefore revisit our previous set of metrics, and add those that would create a more broadly defined set of metrics for our Orchestral musician. For example, we could add the following metrics:

Mentors junior members of the orchestra
Attended two personal development workshops this year
Has developed additional skill on instrument

Made recommendations regarding staffing for section of the orchestra

It is crucial that we are clear in our communications to the associate exactly what expectations there are for exceptional performance within the organization. It is easy for us to attach only those metrics associated with specific deliverables when creating an appropriate appraisal system, but upon reflection we can see the benefit of considering many of the peripheral skills necessary to fully support the organizational goals.

Who should evaluate?

Evaluation is a very difficult task for the manager to accomplish. How can we simplify the task?

Make as many metrics as measurable and objective as possible, and make certain the manager has ample opportunity to observe, in order to make appropriate assessments of performance. Several individuals are typically selected for evaluation purposes, and these typically include:

The immediate supervisor
Peers
Self
Immediate subordinates
Customers

We must consider the appropriate validity of the observations presented by each of the individuals, and combine those when appropriate to increase the reliability and incremental validity of the performance assessment. Therefore, the manager must consider whether there has been ample opportunity to observe the performance characteristics of the associate. In the event he believes he does not have appropriate first-hand knowledge, he should rely upon the input of members of this group.

One limitation should be placed on any of the feedback criticisms included in the performance appraisal associate. That limitation is that any feedback that is provided to the manager must be accompanied with evidence to that opinion. For example, it is inappropriate for feedback to be constructed in a manner as follows:

John is an excellent employee and is a real asset to our department.

Instead, the statement must be accompanied by appropriate evidence to the behaviors, Such as this:

John is an excellent employee and is a real asset to our department, as evidenced by his completion of the Brinkley project last year on schedule and on time. In addition, his project team has recognized him for his collaborative skills in managing the project. Since that time, several other functional departments asked if he would be willing to run projects for them.

How to evaluate

There are many methods utilized by organizations to accomplish personnel assessment. In general, the Engineering leader will be limited by those methods as selected by the Human Resource department. However, consideration should be given to the effectiveness of the selected tool, with an eye towards perhaps making a recommendation for a tool of choice in the assessment process. We will examine several assessment approaches and their attributes.

Essay Method

In this approach, the manager writes an essay relating to the effectiveness of the associate during the performance period. This method is somewhat problematic in that it has been demonstrated that those associates who have managers with better writing skills will typically receive in higher ratings. Due to this, this method has been disappearing within most organizations. Additionally, these ratings are not directly comparable from associate to associate and are considered extremely subjective.

Critical Incidents

In this method, specific critical incidents that are recorded during the performance period are used as a basis to extrapolate the performance of the individual throughout the entire performance period. This method is normally used in organizations where it is extremely difficult to have an opportunity to observe the performance behavior, such as with police officers. Shortcomings of this method are obvious, as these critical incidents are not necessarily representative of the behavior of the individual.

Graphical Rating Scales

Graphical rating scales have been used for a long period of time in performance appraisal. These are typically Likert scales, rating individuals from 1 to 5 with regard to various elements of the performance criteria. One significant difficulty with this system is the typical reluctance to utilize the entire rating scale when completing a survey of this type. As an example, many managers will not use the highest rating, assuming there is always room for improvement. Likewise, they are unlikely to use the lowest rating, as individuals are clearly not as bad as they could potentially be. By virtue of this, most rating scales of this type are artificially reduced from a 5-point scale to a 3- or possibly 2-point scale. Additionally, it is difficult to compare different managers' ratings across associates.

Behaviorally Anchored Rating Scales– BARS

These are relatively common in most organizations, and they are similar to the Likert scale, with the significant difference that each of the rating levels is defined by a behavioral description. As an example:

Five – completed the project as planned, with a twenty percent schedule challenge
Four – completed the project within a reasonable period of time after the scheduled end date
Three – was able to complete the project with significant cost and schedule overruns
Two – was unable to complete the project successfully
One –did not demonstrate effective project management skills, did not inform manager

This anchoring of these skills allows these ratings to be directly comparable across managers and associates within the organization.

Multi-rater comparisons

These comparisons are recommended in any situation where the manager does not have adequate time to observe the behavior of the employee. The intention is to provide further insight into evidence available to other parties within the organization that are pertinent to the rating being conducted. This provides a more representative evaluation for the employee, as multiple viewpoints are considered in the rating. Most effective organizations are utilizing this method in their performance appraisal systems.

There are many challenges with accomplishing performance appraisal in an effective and representative manner for the associates. These challenges include but are not limited to:

Small sample size
Using only a single criterion
Leniency error
Halo or horns error
Similarity error
Having low differentiation between associates

Appraisals designed to match preordained conclusions

Therefore, in order to minimize these challenges and enhance our capability in evaluation, we should follow a simple set of rules in conducting our performance appraisals.

1. Plan what to measure
2. Opportunity to observe
3. Use multiple criteria
4. Deemphasize traits
5. Emphasize behaviors
6. Record those behaviors
7. Use multiple evaluators and raters
8. Appraise only what you know about and have observed
9. Train your people
10. Use performance and continuous feedback

How to Reward

We must give careful consideration to the attractiveness of the reward to the particular individual when we are defining the reward system. Rewards can take many forms, including:

Money
Time off
Recognition
Attractive work
Positive feedback in front of peers
Status
Titles

Many other rewards might be considered appropriate, as well. It is imperative for the Engineering leader to spend some time in reflection regarding the identification of rewards that are attractive to various associates within the organization. Thus, the manager may discover that some associates greatly value momentary rewards, while others will bypass those monetary rewards in order to achieve more attractive work assignments or title advancement. Without appropriate consideration, Engineering leaders may inadvertently select rewards for employees that are not valued and therefore will not act as appropriate motivators within the workplace.

Why is it important to do this well in the workplace? First and foremost, this provides direct feedback to associates regarding their performance. However, in many organizations this appraisal process has been watered down to the point that no practical feedback is provided to the associate. This case is illustrated in Figure 95.

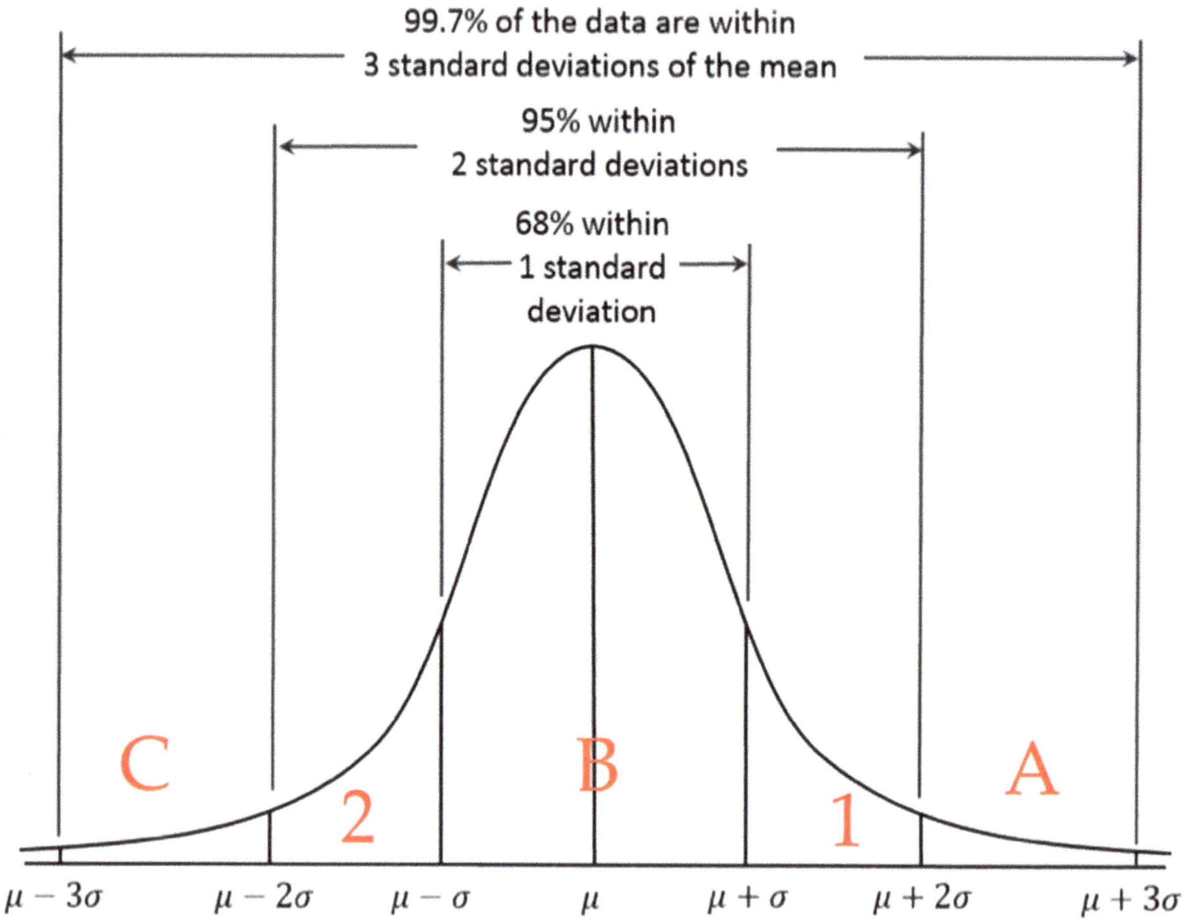

Figure 95: Standard Rating Approach

In this particular case, which is very common commercial enterprises, the rating system has been reduced to a three-point scale. In order to simplify, we are utilizing an A-C system as might be encountered in a university setting. This arrangement results in approximately fifteen percent of the population receiving an A grade, and fifteen percent of the population receiving a C grade. All other members of the population received a B grade and constitute approximately seventy percent of the population. This results in the majority of the organization receiving the same feedback as well as the same financial awards. Imagine two employees, designated by the 1 and 2 of the diagram. Individual 2 is likely of the opinion that her performance is roughly equivalent to individual 1's, as he received the same rating. This of course could be someone in the 16th percentile believing she is equivalent to someone in the 84th percentile. Not communicating this effectively results in person 2 believing she is doing well, while person 1 believes he is not being recognized for his significant work beyond that of person 2. In both cases the performance appraisal accomplishes exactly the opposite of what is intended.

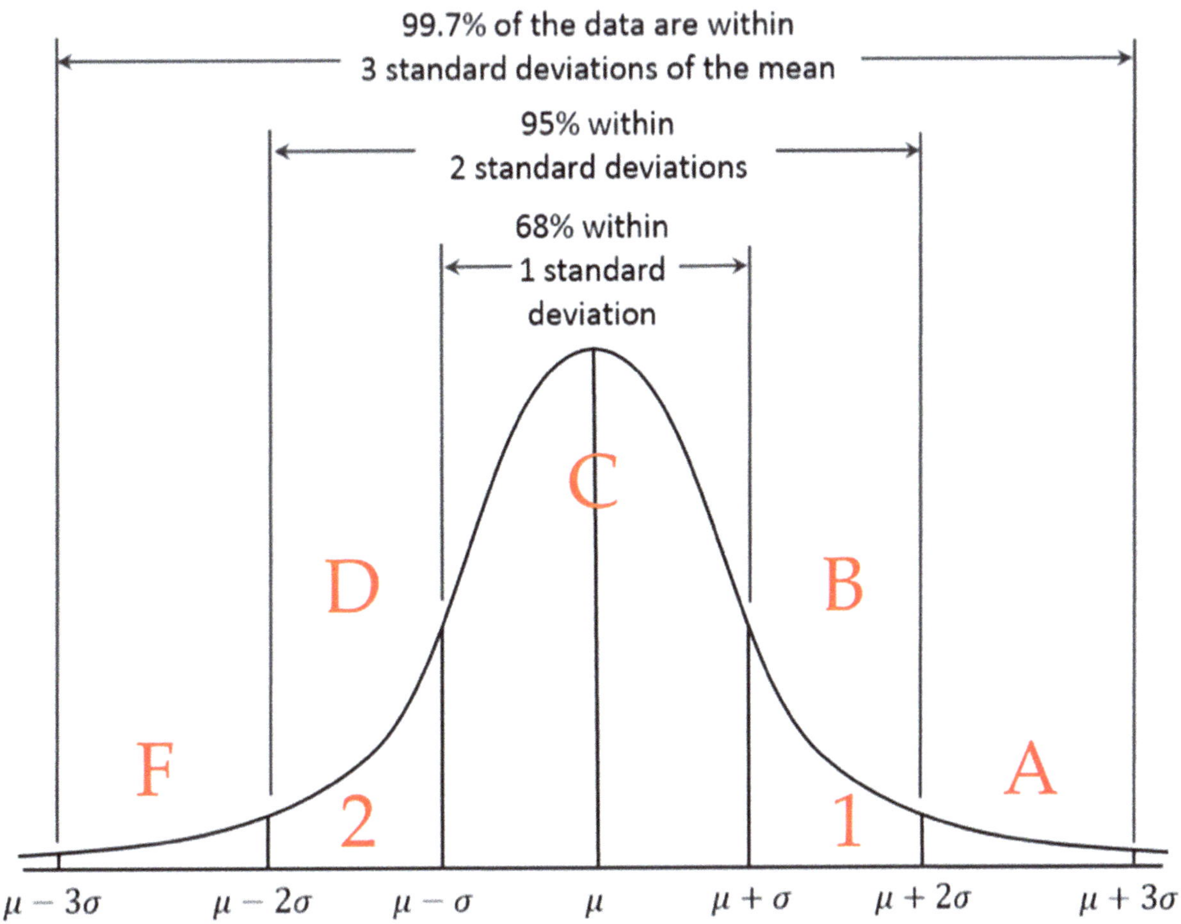

Figure 96: Forced Distribution Approach

In the case of forced distribution, popularized by General Electric, we find that the three-tiered system has been replaced by five tiers. These tiers are represented by the five grades typically encountered in the university setting. In this case, approximately ten percent received an A grade, fifteen percent receive a B grade, fifty percent receive a C grade, and the remaining twenty-five percent received a D or F grade. Rewards are commensurate with the system, with the highest grades receiving the largest rewards, and the D performers receiving no reward. Those individuals receiving an F are separated from the organization. Imagine the difference in attitudes created in the two associates discussed in the previous example. This rating system tends to deliver the message intended by the manager.

In order to accomplish this type of distribution in an Engineering organization, the leaders of the organization must create a pairwise list that includes all of the names of the associates in the organization. At the top of the list is the most highly performing individual, and the bottom of the list is the lowest performing individual. As long as the sample size is large enough to expect a normal distribution to be formed (Approximately one hundred names), the rating is established based upon the percentage distributions in each of the grading criteria. Thus, as long as the leadership gets the list right, the ratings by definition are correct. When relating the performance appraisal to the associate, both the rating and the approximate percentile ranking can be communicated, providing much more relevant feedback on performance.

So why do we find ourselves in a situation where we are required to produce a forced distribution instead of a standard rating as was previously accomplished in our organizations? Largely, this is due to the fact that many inexperienced or unfocused managers chose not to deliver negative news to their direct reports. Because of this reluctance, we created a culture in which mediocrity was typically rewarded in the same manner as high performance. Due to this cultural shift, people became complacent and chose to work at a level commensurate with their particular desires as opposed to focusing on the well being of the organization. As challenges to corporations increased due to economic pressures, it became a strategic imperative to clear out the mediocrity in our organizations in order to remain competitive. Thus, when General Electric published its approach to personnel management, it was recognized as a method by which this could be readily accomplished. In reality, this approach is effective for a limited period of time, after which the organization has a tendency to be separating from medium to medium-high performers in the organization. This is of course a competitive mistake, and most organizations recommend that those individuals in the lowest ratings receive coaching and mentoring first as opposed to separation from the organization. Those that will not or cannot demonstrate personal development are still moved out of the organization.

Chapter 10: Business Rhythm and Reporting

The Project Communications Plan

Project Management communications are crucial as they create the link between the project and the organization. To effectively communicate using written communications requires a systematic process. This process will provide the critical links among the people, ideas, and information that are necessary for Project Management success. Everyone involved in the project must be prepared to send and receive communications, and must understand how the communications in which they are involved as individuals impact the project as a whole.

There are typically three processes required to effectively communicate project status. They include:

Communications planning
Information distribution
Performance reporting

These processes will interact with each other and with the other processes and other areas of the project, as well. Although we may discuss the processes as discrete elements with well-defined interfaces, and practice they in fact overlap and interact in ways that cannot always be understood or easily detailed.

Project communications will differ significantly from standard communication models. Communication models will include issues of sender -- receiver models, feedback loops, and barriers, as well as the choice of media and the selection of how to communicate. Writing style, which might include active versus passive voice, sentence structure, choice of wording and the actual presentation of this information, including body language and design of presentation aids will all impact the content and acceptance of the communication transaction. These issues extend to beyond the scope of this chapter.

Communications planning

Communications planning involves determining the information and communications needs of these stakeholders, which include what information, when it will be needed, how will be given to them, and by whom. Identifying the information needs of the stakeholders and determining a suitable means of meeting those needs is an important factor for project success.

Communications requirements are the sum of the information requirements of the project stakeholders. They might include:

Project organization and stakeholders' possible relationships
Disciplines, departments, and specialties
Logistics of how many individuals will be involved with the project and at which locations and at which times
External informational needs

Information distribution

Information distribution involves making needed information available the project stakeholders in the timely manner. It includes implementing the communications plan, as well as responding to unexpected requests for information. These communications may include work results, and revisions to the project plan.

Several issues will play a significant role in the effectiveness of information distribution. These will include communications skills:

Written and oral, listening and speaking
Internal to the project and external to the project
Formal, in the form of reports and briefings, and in formal such as memos and conversations
Vertical communications and horizontal communications

Information retrieval systems

Information can be shared by team members and stakeholders through a variety of methods, including most effectively electronic databases, Project Management software, and database systems that allow access to the technical documentation of the project. Project information may be distributed using a variety of methods including project meetings, hardcopy document distribution, shared access to network electronic databases, fax, e-mail, voicemail, videoconferencing, and project updates.

Fortunately for the Project Manager, the use of the effective information distribution creates a valuable project record. This project record can later be used in both project reporting and presentations related to the project. In fact, most Project Managers will find the project records invaluable in assessing the post-mortem analysis of the project.

Performance reporting

Performance reporting involves collecting in disseminating performance information to provide stakeholders with information about the status of the project. This process includes:

Status reporting, which will describe where the project now stands

Progress reporting, which will describe what the project team has accomplished during the previous period.

Forecasting; that is to say, predicting the outcome of the project and expected progress

Performance reporting should generally provide accurate information on specification, schedule, and budget. Some projects will also require incidental information such as risk management and procurement status.

Many tools and techniques exist for accurate performance reporting. These tools may include performance reviews, variance analysis, trend analysis, and most importantly, earned value analysis. The effective use of earned value analysis will provide the Project Manager with an accurate assessment of the status of the project with respect to both the efficiency of the project team, as well as the spending rate of the project.

Business Rhythm

Business Rhythm is the term coined to establish a regular set of recurring meetings that are defined in a manner that assures updates cascade (upward) throughout the organization. The primary goal of establishing an effective business rhythm is to assure ourselves that the information presented at periodic updates is both relevant and fresh.

In many large organizations, considerable time is spent in generating and providing updates to higher levels of the organization. In fact, in many organizations, if the true cost of these updates were calculated, many leaders would rethink their policy of demanding these regular and detailed updates. In any case, the use of a business rhythm can dramatically reduce the workload on the associates providing updates by establishing a tiered method of updating for all individuals.

As an example, to establish a business rhythm within a large manufacturing company; each of the associates within the company is provided the opportunity to attend a weekly staff meeting. At the weekly staff meeting, associates have the opportunity to update their managers on the current statuses and outlooks of their projects and tasks. A second meeting is scheduled for the First-line Managers following the individual staff meetings. A monthly meeting is established for the Senior Manager level, where all of the functional projects are updated based upon information received at the lower levels. This monthly meeting is scheduled to occur just prior to the quarterly update meetings that occur throughout the year.

Many different constructions can be defined for an appropriate business rhythm, and consideration should be given to the reduction of materials that are generated and presented numerous times within the organization. By staggering the updates from the bottom of the organization to the top, many of these materials can be consistently reused for higher-level status updates. An example business rhythm schedule is shown in Figure 97.

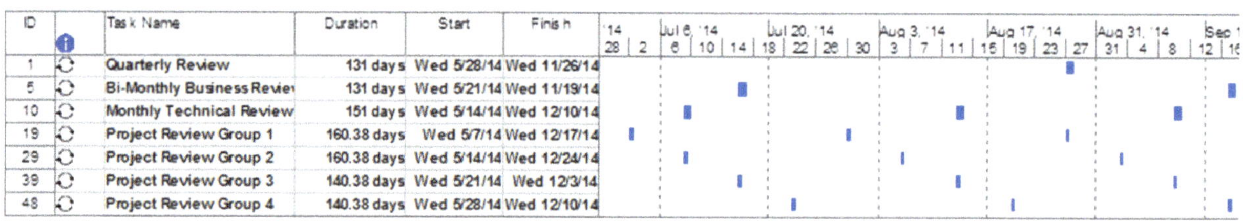

Figure 97: Business Rhythm Schedule

Project Meetings and Status

Meetings should be planned periodically to ascertain status and predict outcome.

1. Set an agenda for project meetings to review progress
2. Define action points against each item on the agenda
3. Review the items on the critical path

4. Report if the cost or time limit exceed plan
5. Report progress at the end of each stage of the project
6. Monitor issues that may be causing concern

Project Reviews

Project Status will provide timely information on the current status of the project, but the use of Project Reviews is recommended to assure the project achieves all objectives and satisfies the customer.

Project Review Key Objectives:

Technical review of system and / or sub-systems to confirm they meet Functionality, Robustness, Specifications and Customer Needs

Cross-disciplinary review to ensure optimized ***system*** solution

Documentation to comply with Product Development Process and support Design History File

Results of conducting project reviews:

Catch mistakes at the design stage, different eyes looking at it

Save money on debug, iterations, expedite fees (payments)

Product meets specifications and customer needs

Team coherence on design and a better product

Participants for a Successful Review

Who should be included in design reviews and does it depend on the stage of development?

All stakeholders in the project as a starting point

Production
Marketing (represents Customer)
Regulatory / Quality
Project / Program management

Always invite all representatives. Require attendance based on stage of development and topic of review. The team owns the design and is responsible for successful completion, so critiques are of value and should be encouraged.

Running a Successful Review

Materials – distribute them before the review:

Relevant Specifications and Requirements
Detailed design drawings, schematics, simulations, flow charts, test plans, etc.
Known issues, discrepancies with specifications, conflicts with other disciplines

Agenda with timeline by discipline
Structure – review the project in its entirety by discipline and subsystem
Start on time and stay on schedule (Project or Engineering Manager leads)
Not a formal presentation, rather a discussion approach, so encourage interruptions
Focus on improving the design, no personal attacks
Action items defined for changes, items to check, functionality changes, etc.
Decisions about design approach, specification changes, trade-offs, etc. should be made
Meeting minutes including attendance, references to materials covered, reference to actions / issues / decisions list
Need documented closure on all actions before completing the next design step, this becomes a gate
Add to Design History File (including copies of materials reviewed)

Concept Design Review (Preliminary Design Review)

Concept review / update
System level architecture is the focus
Identification of high-risk areas of design with focus on technology challenges
Modeling / simulations / breadboard results demonstrating risk mitigation for high risk areas of design
Goal: To provide approval to begin detailed design work

Participants

All appropriate team members
Team members from related projects in the program
Balance group size with required participation

Materials

Hardcopy of agenda, all drawings, schematics, specs, modeling and simulation results
Physical prototypes if available
All materials distributed prior to the meeting

Review Execution

1-2 hours (good planning to optimize resource time commitment)
Can break into multiple reviews distributed over days / weeks
Clearly stated goals and objectives with list of materials to be reviewed
Examine if the feasibility is appropriate, and if the approach (with data to support) makes sense to those involved

Be prepared to articulate risks and mitigations based on proof of concept prototype activities (linked to specifications, traceability)
Project Manager leads review
Meeting notes required with Actions / Issues / Decisions
All materials and notes to Design History File

Detailed Design Review (Critical Design Review)
Participants: Same as previous review
Materials: Same as previous review plus detailed work accomplished
Goal: To provide approval for design release for production

Review Execution
2+ hours (good planning to optimize resource time commitment)
Clearly stated goals and objectives with list of materials to be reviewed
Highly technical, quantitative review of **all** aspects of design, all design drawings, all software, etc.
Be prepared to articulate detailed analysis of designs, decisions, trade-offs
Review of specifications (line-by-line) to ensure design completion and adherence
Completed review should give everyone the feeling that this is an imminent success and release for build

Project Close-out - The Project Post-Mortem

With all the challenges that Project Managers face in the development of complex projects, it isn't particularly surprising that very few of us take time to conduct project post-mortems. The function of a post-mortem is to summarize a given project history and analyze all the positive and negative impacts on the project in hopes of developing a predictive view of how our projects will function in the future. Essentially, a Project Manager is assessing his or her own ability to predict the future and then perform to that prediction. This assessment will often result in new guidelines or recommendations with regard to Project Management approaches going forward.

An effective post-mortem is typically conducted within several weeks of completion of the project. This allows the individuals conducting and participating in the post-mortem to have fresh memories of the events of the project, and will be separated sufficiently from the project closure that emotional input can be minimized.

Here are ten suggestions for conducting a good project postmortem:

1. Involve all the contributors. If others were involved in your project, arrange a project review. If you were the sole contributor, set aside time to reflect on your experience. Document the good, the bad, and the ugly aspects of the project. Note all grievances, and solicit ideas for future improvement.

2. Document the post-mortem in writing. The document can then be published so everyone in the company has easy access to it. A series of post-mortems can ultimately evolve into a valuable record of your expected performance.

3. Begin with a project overview. Begin the post-mortem document with a brief overview of your project, including a design overview, estimated budget, and project start and finish dates. Describe the product's purpose, intended customer, and other general information.

4. Include project details. Document the quantifiable details of your project. Include your schedule and budget baseline as well as the true outcomes. List everyone who worked on the project, and document his or her contributions.

5. Document what went right. Document what went right with your project. Did the final product turn out better than expected? Did you experience an occasional stroke of luck? List at least ten things that turned out well for your project.

6. Document what went wrong. What difficulties did you encounter? What assumptions proved to be incorrect? Compile a detailed list of your project's shortcomings. List at least ten things that turned out worse than expected. Document unpleasant mid-project surprises, as well.

7. Assess your risk management. Document all the risks you took during the project, and note how effectively you managed them. Did you explore experimental ideas, use new technology or development tools, or develop for a new platform? Did your schedule slip from poor estimation practices? Did you take big risks, or did you play it relatively safe? Assess your risk management experience, paying particular attention to how you would change your approach for the next project.

8. Assess your cost and schedule performance. How did you perform on schedule variance and cost variance? Using earned value analysis can often provide excellent predictive statistics for use on your next project. The same statistics can often be used as corrective factors on other people's projects as well.

9. Draw meaningful conclusions. What conclusions can you draw from your project history that will help you improve in the future? What should you start doing, keep doing, or stop doing? Describe new practices you should try for your next project. This is perhaps the most important part of the post-mortem, so be prepared to spend more time here than in any other section.

10. Take action. The final step is to develop an action plan that can be applied to your next project. Review your project history, reflect on the lessons you've learned, and specify new guidelines to follow in the future. Create a checklist for your next project. If you already have a checklist, then update it with new refinements. One final benefit of conducting a post-mortem is that you gain a very gratifying sense of closure because you actually document all the personal growth you've experienced as a result of completing the project.

So what should be on the list of questions? I've provided some of my favorites below.

General Questions

1. Are you satisfied with the results of the project?
2. What was most frustrating about the project?
3. What was the best part of the project?
4. How should we do things differently next time?
5. What methods do you like the best in our project development?
6. What methods do you like the least in our project development?
7. What three things would you change on this project if you were to start over today?
8. How did our sponsors do?

Project Initiation

1. Did we understand the strategy of the project?
2. Did we understand our alignment with the corporate strategy?
3. Did we identify all the general deliverables on the project effectively?
4. Did we make assumptions on deliverables that we should not have made?
5. Did we get the team engaged before we begin the planning process?

Project Planning

1. How accurate was our original estimate?

2. How could we have improved our estimation process?
3. Does our original specification describe what we ultimately accomplished?
4. What could we do (or not do) that would improve our specification in the future?
5. Did we complete a network diagram describing the inter relationships of the functions of the project?
6. Did our work breakdown structure describe what was accomplished on the project?
7. Did we have the right people assigned to the project roles?
8. How can we make sure we get the right people assigned next time?
9. Did we recognize any problems during the planning phase that occurred later in the project?
10. Could we have completed this project in less time for less cost given what we know today? How can we incorporate these lessons into our future project?
11. Were all of the assumptions made clear to everyone on the project?
12. Did we go down some wrong paths?
13. Did we negotiate effectively during the project-planning phase?
14. Was there anyone missing from the project kickoff who should have been involved from its outset?
15. Were the roles and responsibilities for all of the team members and stakeholders clearly defined and communicated?
16. Were the specifications, milestones, schedule, budget, user needs, deliverables all clearly communicated?

Project Execution

1. What was the final quality of our deliverables? Could we improve on this?
2. Did the project culminate with innovative and creative solutions?
3. Did everyone have input into the project outcome? Did we engage all the players?
4. Did we have effective project and reviews?
5. Did we receive critical inputs from outside parties positively? What can we do to improve this?
6. Did we treat our team members with courtesy and respect?
7. Does the team want to work on your next project? Why? Why not?

Project Control

1. Did we do a good job overall?
2. What could we have improved on this project?
3. Were we able to ascertain the status of the project at any given time?
4. Were our project predictions accurate?
5. Were we able to do what we said we would do?

6. How can we take the performance on this project to predict performance on our next project?
7. What methods or processes helped you better understand the status of the project and which did not?
8. If our project deviated from the specified path, were we able to bring it back in line with expectations?
9. Did we manage expectations effectively?
10. Did we succeed? Why? Why not?

Project Close Out

1. Did we have the right members on the team?
2. Did we have the right customers in user needs?
3. How could we do better in the future?
4. Did we understand our customer?
5. Were our assumptions representative of our customer's needs?
6. Did our project ultimately meet our customers' needs?
7. Did we miss the mark in any way?
8. Were we able to manage our project effectively during the execution and control phases?
9. Did we get timely feedback on design reviews and oversight from our sponsors?
10. How did the sponsors do overall?
11. How did the team perform?
12. What have we learned about the team and what would we do differently?
13. Was our original strategy effective?
14. Is the customer coming back? Why? Why not?
15. How will the lessons learned on this project be communicated to the next?

Chapter 11: Earned Value Management Systems

In order to realistically manage a project or projects in an enterprise, it is beneficial to understand a practical and relevant manner in which to status a project. Many systems exist within organizations in order to accomplish status definition. In general, Earned Value Analysis provides significant predictive capability for the Project Manager or Engineering leader, and is therefore a preferred method.

Earned Value Management Systems (EVMS) is a cost management technique wherein the value of completed work is compared to actual costs and the scheduled costs for the project. By examination of these two cost ratios, and excellent prediction may be created to define the expected outcome of the project.

Earned Value is typically expressed in terms of dollars, thereby allowing a comparison between completed work, timecards and actual costs on the project.

Earned Value is calculated by multiplying an estimate of the percentage complete for the task by the total estimated cost for that task. As an example:

Five weeklong test costing ten thousand dollars

Task is reported as thirty-seven percent complete

$0.37 \times 10{,}000$ = \$3700 of Earned Value

Therefore, in this instance the Project Manager is allowed to take a \$3700 credit for work that has been completed. This value would then be compared to the actual cost to complete the work, thereby giving some insight into cost. Secondly, this value will be compared to the work scheduled to be completed, thereby giving some insight into variance in the schedule.

So why would Earned Value Analysis be of any interest to the Project Manager? It provides us with a significant ability to predict outcome of the project prior to completion. As an example, consider a \$100,000 project that is currently underway:

We have spent \$90,000

The earned value of the work completed is $80,000

Most people would recognize that this predicts that we will have a significant cost overruns on our project. Earned value allows us to answer the next question to be asked regarding this project; what is the amount of the cost overrun? To establish this, we must think through what this information provides us regarding our current status.

We have $20,000 of work remaining
We have $10,000 of funding remaining
We know we are accomplishing our work at and 8/9 efficiency
Therefore we will require approximately $22,000 to complete the work
Therefore the expected cost overrun is $12,000

And identical analysis can be performed related to the schedule status as well.

Several variables are necessary to establish the earned value of the work. They include:

BCWP – budgeted cost of work performed
ACWP – actual cost of work performed
BCWS– Budgeted cost of work scheduled

To establish metrics for our project, we are interested principally in two factors. These include the cost variance, relating to the efficiency with which we are completing our work, and the schedule variance, relating to the rate at which we are completing our work. With these two pieces of information we are enabled to make an accurate prediction regarding the final outcome of the project once the project is underway.

CVI (Cost Variance Index) = BCWP/ACWP
SVI (Schedule Variance Index) = BCWP/BCWS

Both of these variables are utilized in a similar manner when assessing the status of the project. If we utilize the cost variance index as an example, it is obvious that if the budgeted cost of work performed is equivalent to the actual cost of work performed, we have a CVI equivalent to 1.0. This implies that we are paying as much for the work as we predicted we would pay for the work accomplished. This would lead us to believe that we will finish the project on budget.

When we consider how projects are normally evaluated, we typically see the Gantt chart utilized as the status tool. This is shown in Figure 98.

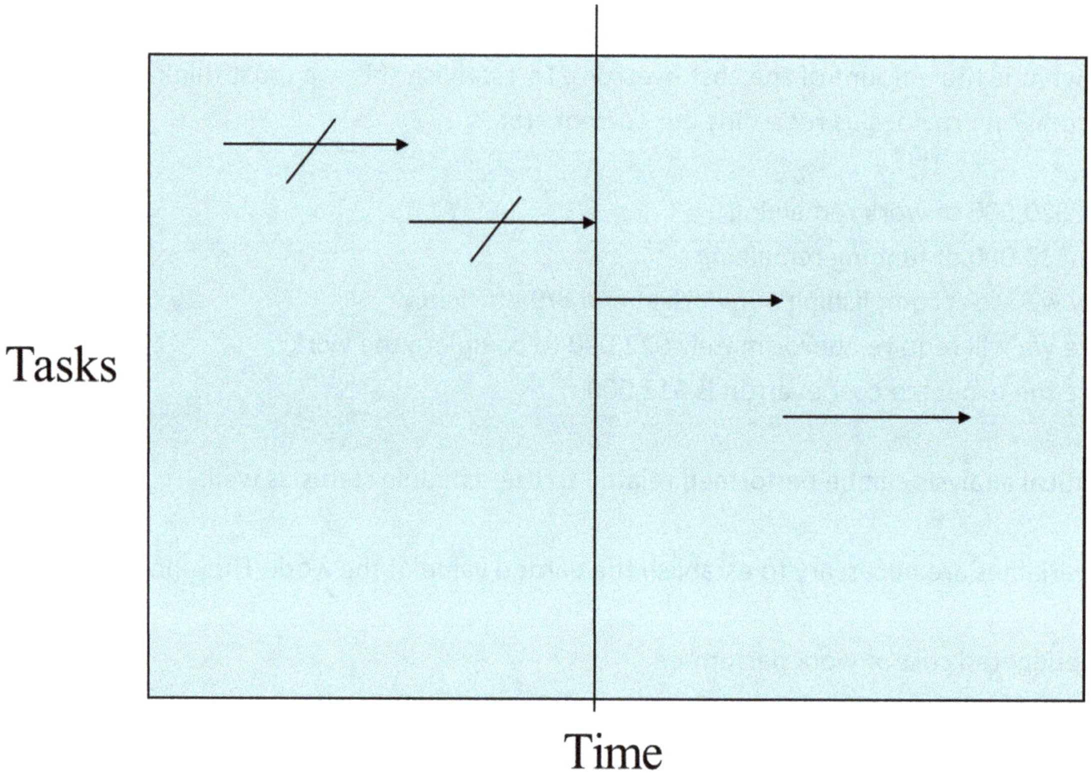

Figure 98: Gantt Chart Status

As can be seen, our tasks are listed on the project plan, two of which have been completed and with a vertical line representing the current date. Most Project Managers would assess this project as being on schedule, but they may not have considered several interesting factors related to the status of the project:

The status was supplied by the individual assigned the task.
It may be of interest to the associate to report work is complete that they fully expect will be completed in the near future.
There is considerable question as to what the definition of 'done' means.

By virtue of these considerations, it is potentially possible that although the project appears to be on schedule utilizing the Gantt chart, in fact the project might be off schedule.

Additionally, if the third task were to be reported as thirty-seven percent complete, we could consider the earned value of that specific task in our analysis. But in the big scheme of things, to the Project Manager, this means one thing: the task is not complete. Therefore, we see that if the tasks are not sufficiently aggregated, any reporting approach may provide us with less than desirable data. In general, it is advised that no task should exceed the amount of error that you can support in your project tracking. Therefore, many Project Managers will set a rule that no task may exceed forty hours in duration, or something similar.

The second most commonly used tracking tool for Project Managers is the current budget. We can make a similar inaccuracy (or non-representative) argument relating to the consumption of the budget during the execution of the project. Remember, without a complete understanding of how the consumption of the budget relates to the completion of the work remaining, it is a dubious metric to be considered.

Rather than use these typical metrics for our project assessment, we instead will use two EVMS metrics in order to make a project prediction.

Is the work getting completed at close to the estimated cost?
Is the work progressing as originally scheduled?
From this we will predict if we can meet our specification, schedule, and budget

Cost variance is a relatively simple concept, and it describes the deviation of actual costs from expected (or budgeted) costs. This can be seen in Figure 99.

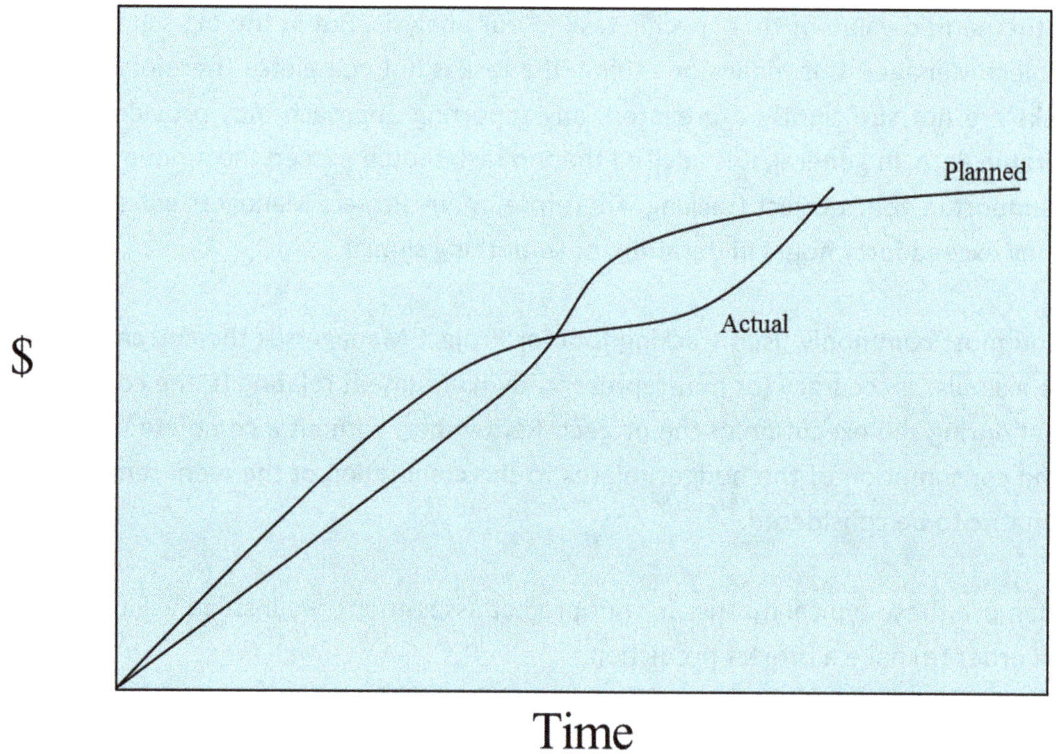

Figure 99: Cost Variance

In this example, the actual costs are shown to have been outpacing the expected costs throughout the first half of the project, and now are significantly less than the expected costs. The real challenge lies in deciding how we would utilize this information to make a prediction of the actual outcome of the project. One can see that many extrapolations of the information could plausibly be used, so which is the most predictive?

Schedule variance is related to the rate at which work is being completed on the project. It is in essence a comparison between the tasks that are completed at this point in time, and what tasks were expected to have been completed at this point in time. The comparison of these two values enables us to gain some insight into the expected outcome of the schedule for the project. This is shown in Figure 100.

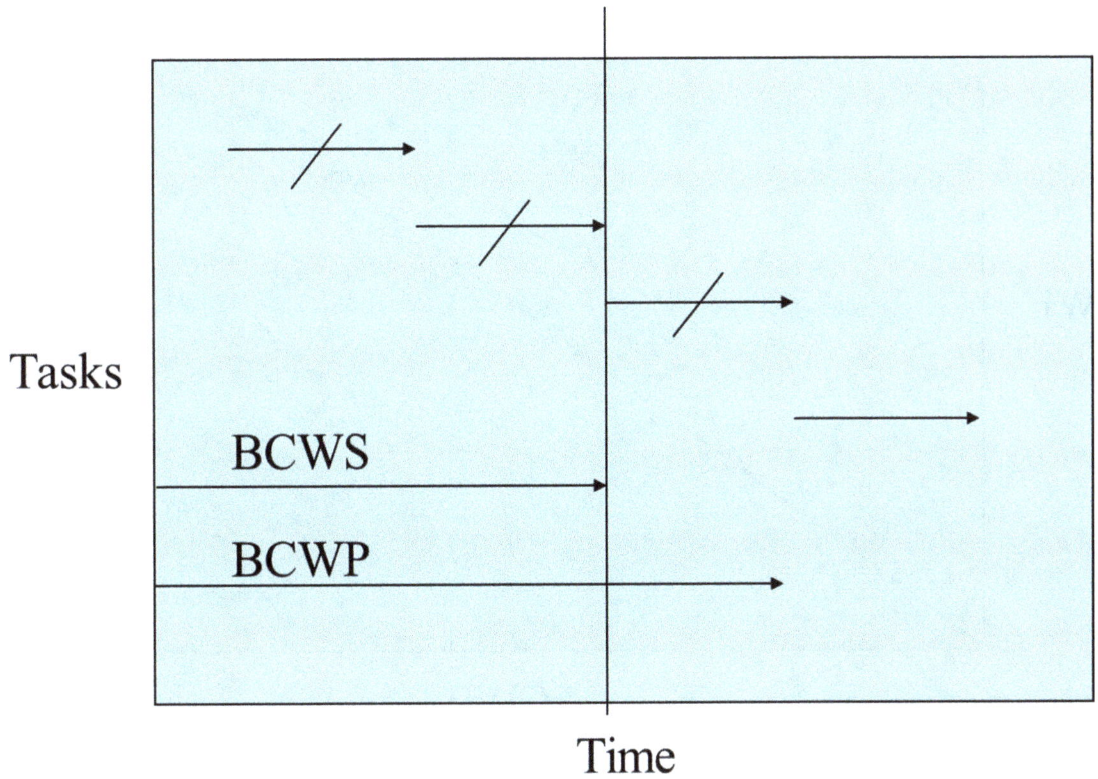

Figure 100: Schedule Variance

These variances can be of significant importance to us, but we must also understand the dynamic nature of the project environment. Regardless of the appropriateness or accuracy of the metrics used in the tracking of a project, due to a wide variety of considerations within the organization the actual path of the project is very likely to deviate significantly from the expected path. Therefore, it is likely that we will continuously be deviating from the planned path during the execution of the project. We must consider these variances with regard to the appropriate control limits as the project progresses. Thus, we will consider that if the project is within a given percentage of the expected position, we will limit our interaction with the project team in order to minimize the potential Chaos interjected into the project environment. The multiple paths that different development teams may follow in the completion of the project and appropriate control limits are shown in Figure 101.

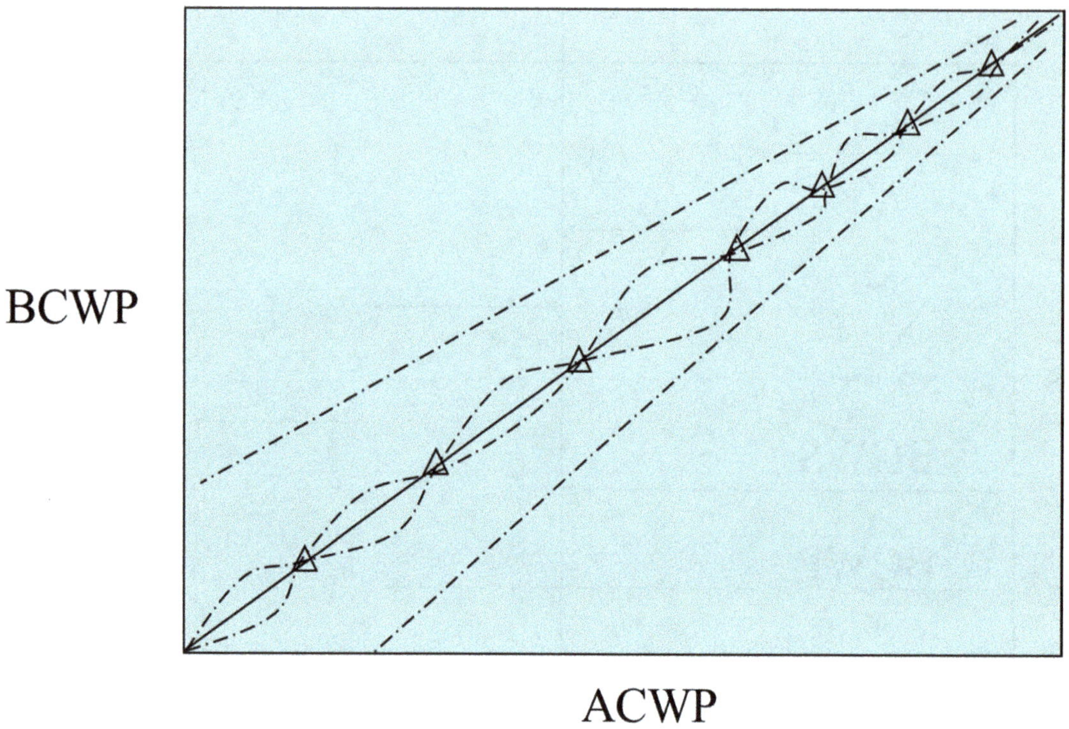

Figure 101: Control Limits for Cost Variance

Therefore, we will consider both the cost and schedule variances in our typical assessment of project status. As an example, in the following diagram we are reporting a cost variance of 1.1 and a schedule variance of 1.0. This is shown in Figure 102.

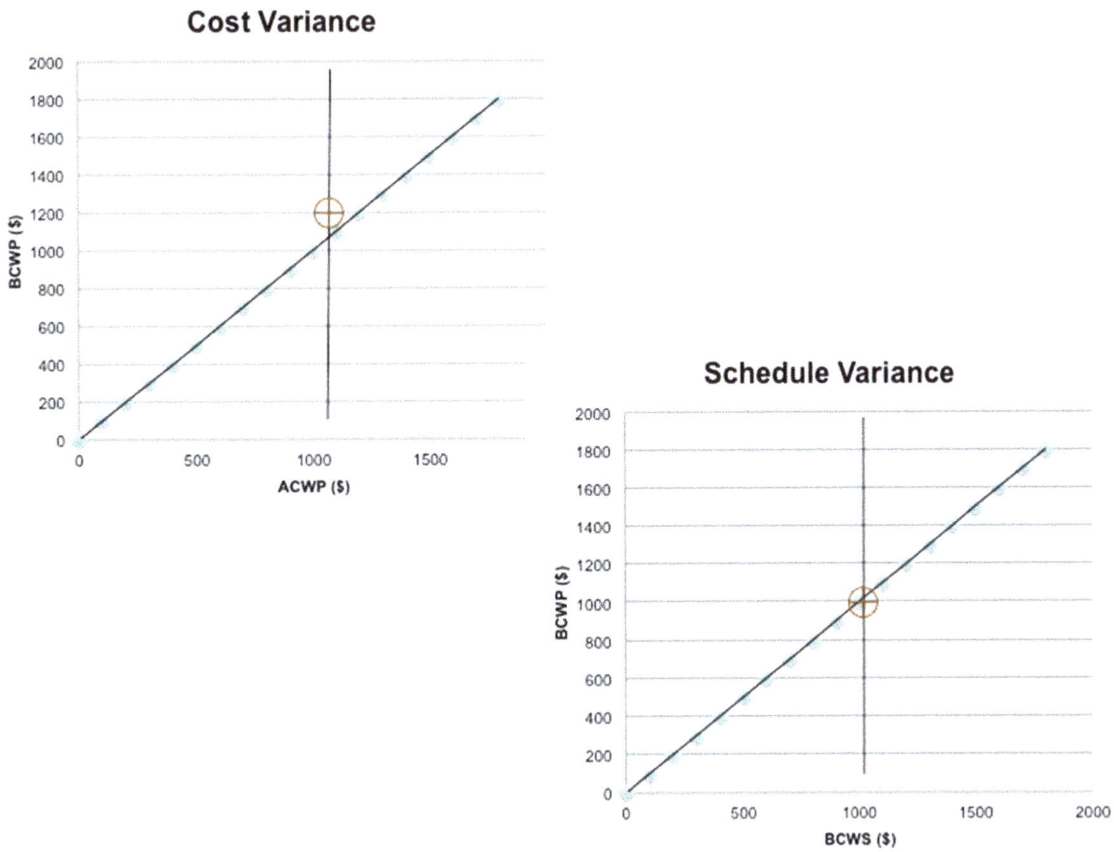

Figure 102: Project Status Charts

From these charts, we can see that the work is progressing as expected, and therefore it is likely we will meet our original schedule. The cost variance has a favorable value, and we would predict that we will deliver this project under budget by approximately ten percent. These results are extremely straightforward and typically are no surprise to the reader.

In a more sophisticated example, we have reported a cost variance of 1.0 and a schedule variance of 0.5. This is shown in Figure 103.

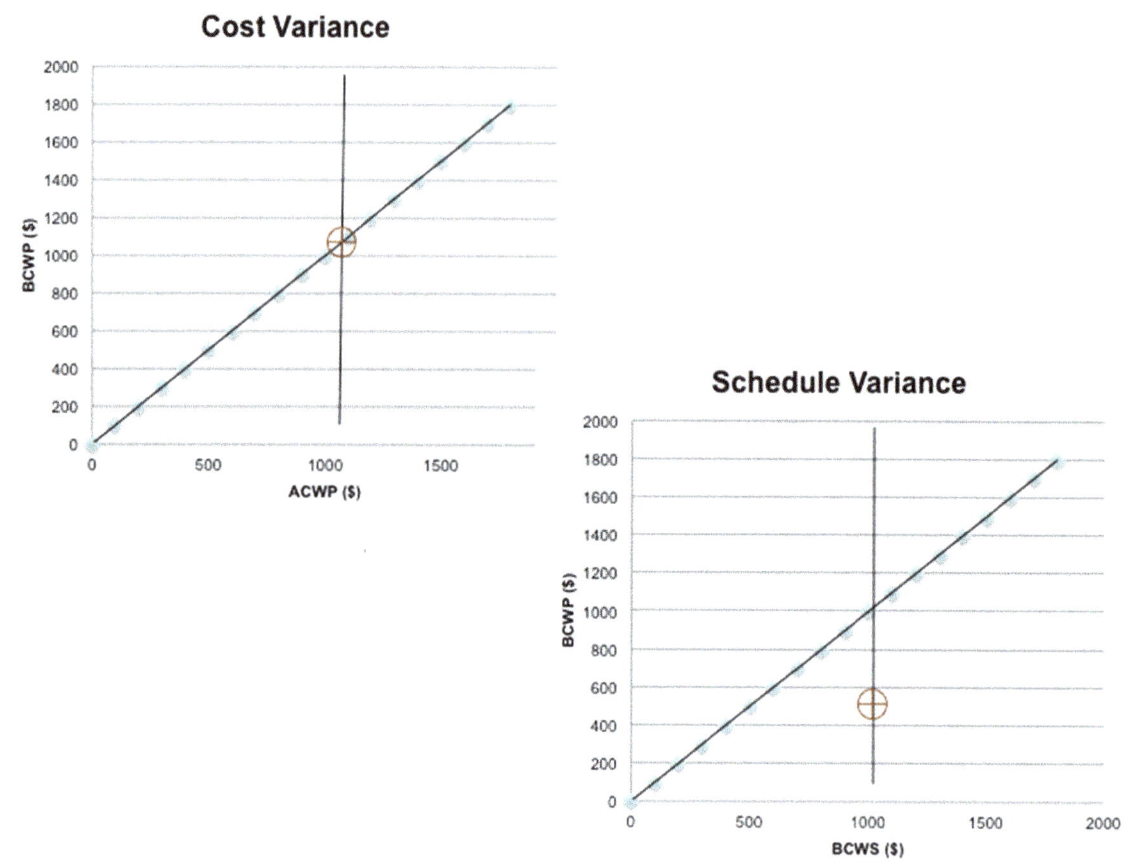

Figure 103: Project Status 2

This project has a much more interesting story to tell. It appears from our cost variance that we will be able to complete our project within the specified budget. However, we have a significant problem relating to our schedule variance, and that is that only half of the expected work is completed at this time. In order to get this project back on schedule, we need to add people to our team in order to accomplish more work at a quicker rate. It is possible to examine the schedule variance and make a prediction of how many personnel need to be added to the project. In this case, we have spent one-half of the time and completed one-quarter of the work. In the remaining one-half of the time, we had three-quarters of the work yet to complete; therefore we must triple our current workforce in order to meet our original deadline.

One concern that is frequently voiced by Project Managers is that the addition of significant personnel to the project team will also cause significant cost overruns. While that may be true due to inefficiencies that are inherent in the expansion of any work team, in this particular case it is not a significant concern. In fact, the adjustment of the workforce in order to adjust the schedule variance will typically have little or no bearing on the cost variance.

As an example, let us assume that we have hired a neighborhood boy to mow our lawn. Normally this requires three hours of effort and he is rewarded with $30 payment. This results in a $10 per hour rate for completion of the job. Imagine that we have scheduled a significant event on our lawn one day, and the individual responsible for mowing the lawn no longer has three hours in order to complete the work. If we request he complete the work in one hour instead of the usual three, (we would like the work completed prior to the event), then he must increase his workforce. In this case, let us assume that he brings along two friends to assist in the mowing of the lawn. As might be expected, the work is completed in one hour. The payment is still $30, which in this case will be paid to three individuals, and still results in $10 per hour as the rate of pay. Therefore, the cost variance has not changed even though the workforce has been tripled in size. This of course ignores the potential problems associated with multiple personnel mowing the lawn simultaneously.

In most organizations, the EVMS information is not expressed in a manner as simple as the previous two diagrams. In fact, it will look more like the data presented in Figure 104.

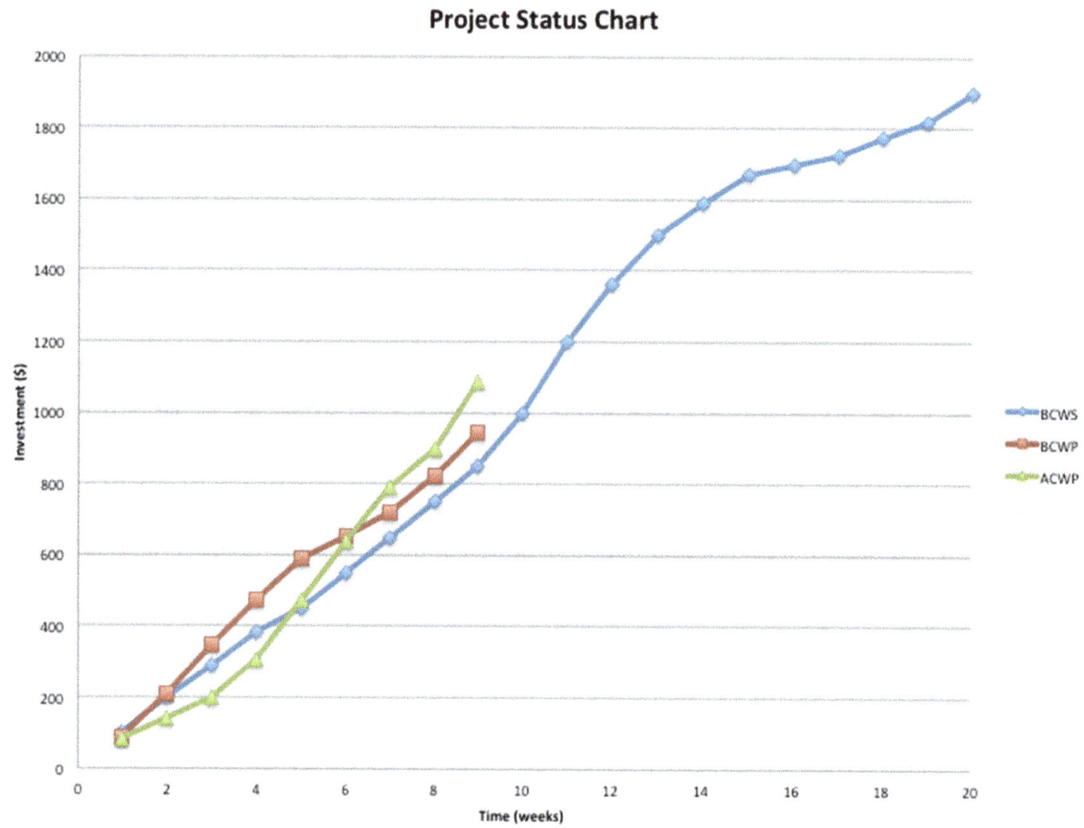

Figure 104: Project Status 3

While the same information is present on the chart, this method of reporting can be quite problematic with regard to predicting the outcome of the project and managing it accordingly. It may be beneficial to reconstruct diagrams from this format to those described in the previous diagrams. This will alleviate the chance of misinterpreting the information on the chart, as it requires several translations of the data in the reader's head. If this is not possible, it is often better to simply report the ratio values rather than a graphical chart.

Regardless of the method selected, the user must utilize this information to decide if action must be taken to refocus efforts on the project. This can be defined by meticulous analysis of the EVMS ratios, but fortunately there is a much better way. This simplified method is to utilize the system of a bull's-eye chart.

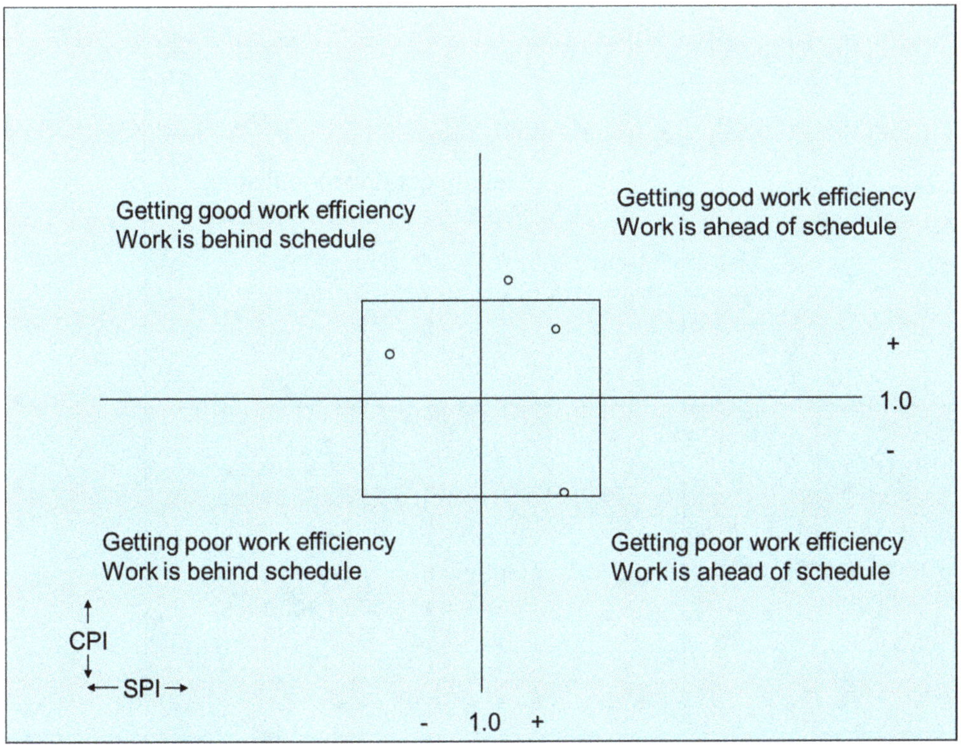

Figure 105: Bulls-eye Chart

In a bull's-eye chart, the cost variance index and schedule variance index are located on the Y-axis and X-axis respectively. As can be seen in Figure 105, this places the ideal performance of the project residing at the bull's-eye in the center of this chart. A control box may also be supplied on this chart as a reference to establish where the desired performance for the project resides. This control box is equivalent to the control limits previously discussed.

Data can then be provided on this bull's-eye chart over a period of time, creating a longitudinal view of the status of the project. This gives the Project Manager a quick insight as to the status of the project, with little danger of misinterpretation. Additionally, criteria for assessment can be included on the bull's-eye chart, providing general recommendations to the Project Manager. This can be seen in Figure 106.

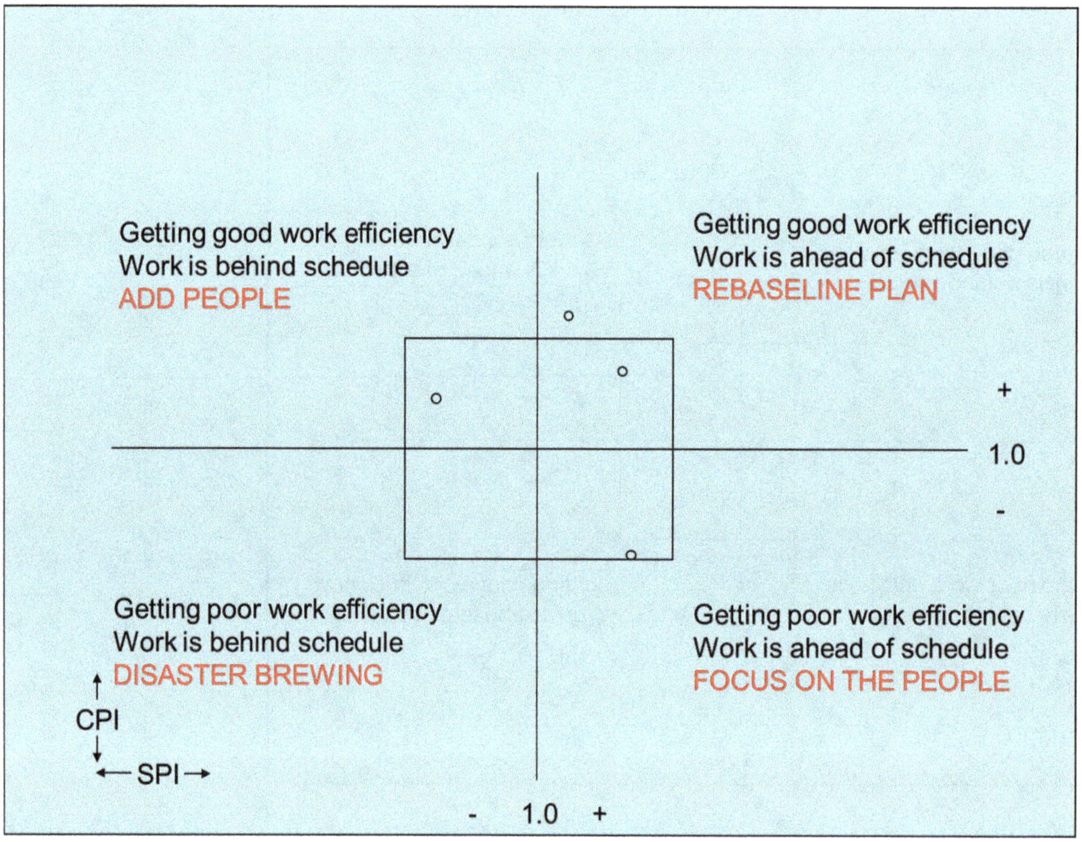

Figure 106: Bulls-eye chart with recommendations

An example of a real bull's-eye chart with twelve months of status data posted is shown in Figure 107. Note the color coding of the chart; showing green, yellow and red zones of performance.

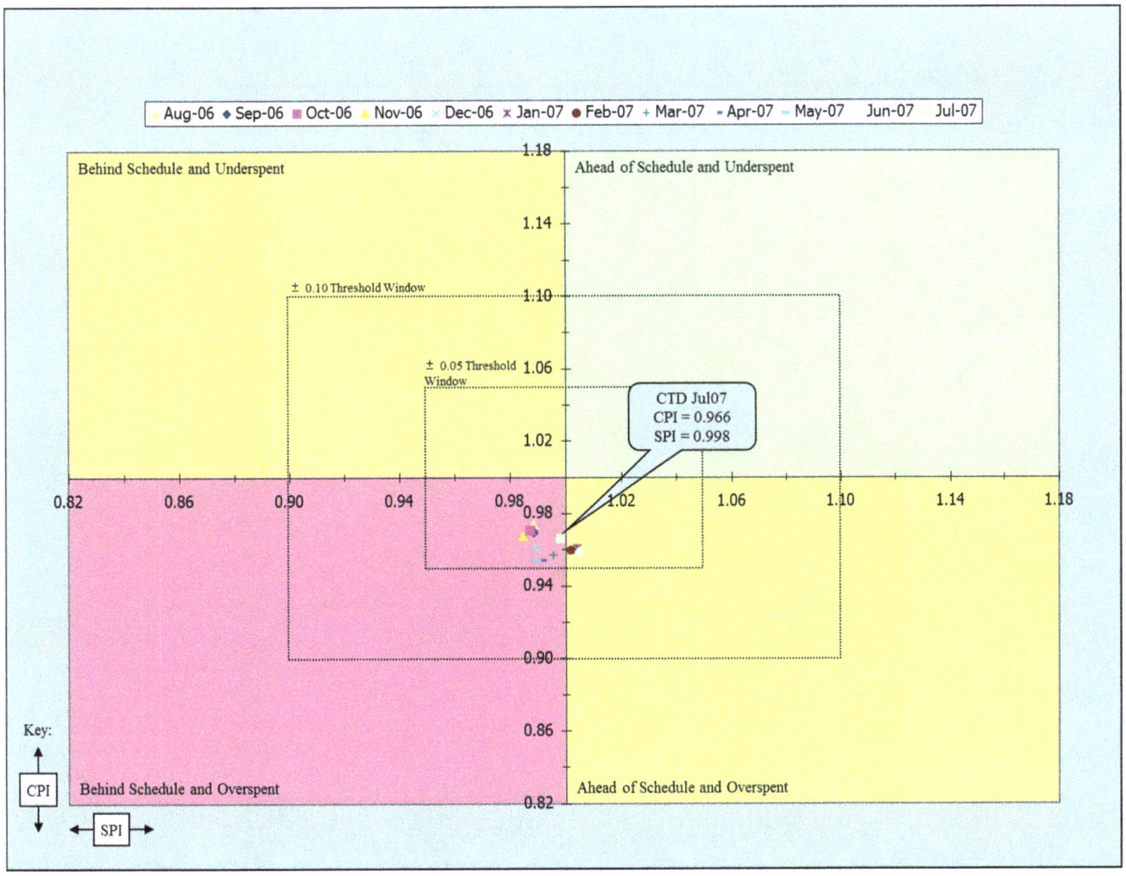

Figure 107: Example Bulls-eye Chart with Twelve Months of Data

Extrapolation of the data provided by the cost variance and schedule variance is often used to make a prediction of project outcome. The methods by which this extrapolation is accomplished are shown in Figure 108. Note that this is using the total project performance to define the predicted outcome, as past performance is indicative of future performance. This is normally the best prediction for a Project Manager to use when extrapolating. One caveat should be mentioned; early in the project the data can be very volatile, so exercise caution when predicting outcome based upon less than twenty percent project completion.

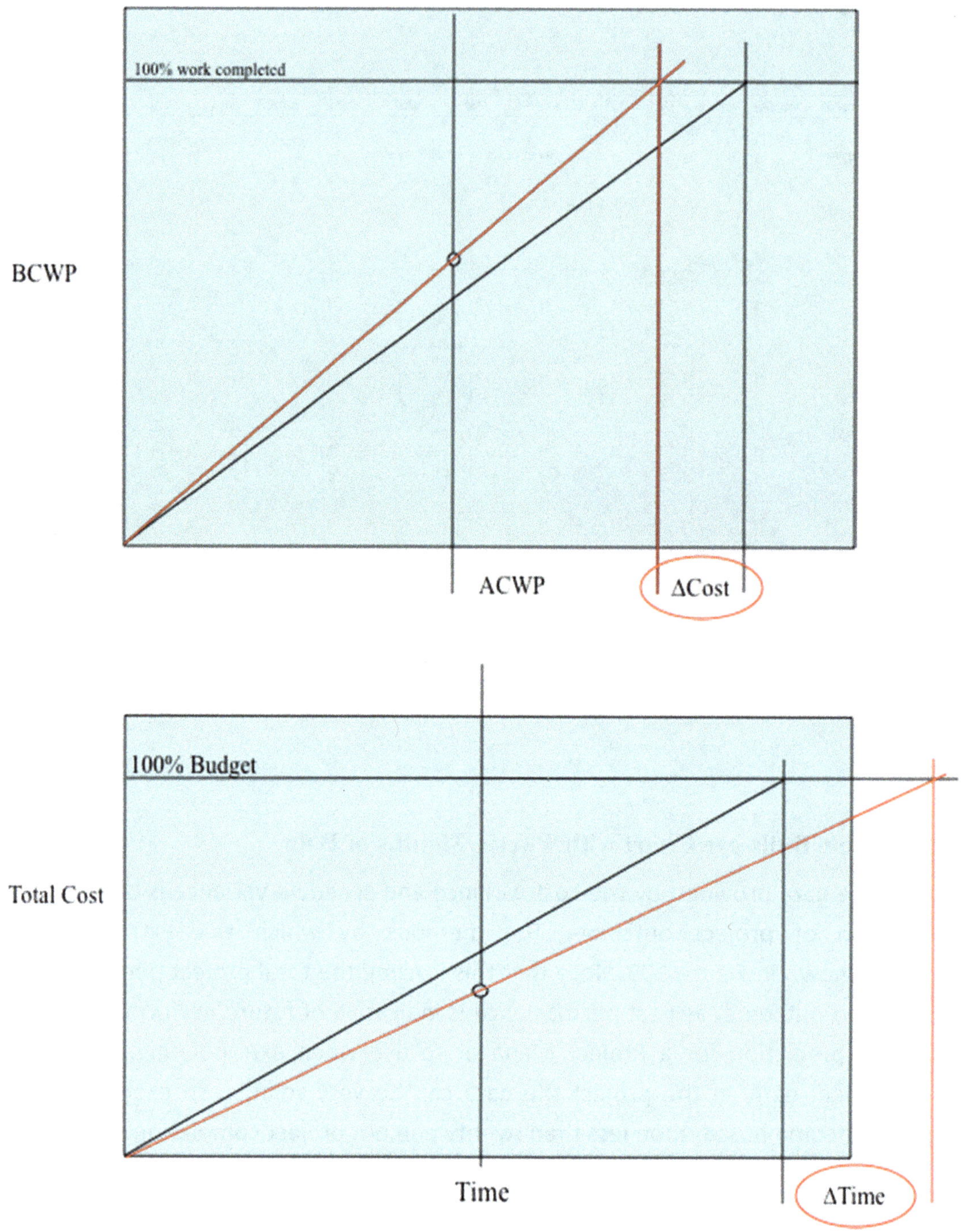

Figure 108: Extrapolation of Project Outcome

Since it is necessary to reach one hundred percent of the budgeted cost of work completed (BCWP) in order to finish the project, we utilize this line as our limit for prediction as shown in these two diagrams.

It is extremely important for any Engineering leader to understand how to quickly assess the status of a project given these two variables and define appropriate actions forward to enable maximum probability of project success. Several examples are provided here for consideration.

Predicting Project Outcomes

Your project is progressing reasonably well, but you are worried about the reports you have received on the earned value of your tasks. You have gotten differing results over the past three weeks, and you are very concerned about what this means to the project. Assuming you are halfway into the project, what do you report for each of the following scenarios?

CVI = 1.1, SVI = 1.0

First off, CVI is greater than one, so we know we are getting more work done than we estimated per dollar; our efficiency is good. Therefore, we know we should finish ahead of budget. The SVI shows we are spending at the right rate, so we should finish on time, as well.

CVI = .85, SVI = 1.2

The CVI says we are not as efficient (our estimates were optimistic) as we planned, so we should now predict the amount of overrun based on a reduced efficiency, a fifteen percent overrun. SVI tells us we are ahead of schedule, which might be a good argument for us to ask for further funding, as it appears we will finish before the projected end date.

CVI = 1.1, SVI = .95

The cost variance is favorable, but the schedule variance is unfavorable. They are both reasonably close to 1.0, so we can probably make up the schedule variance by applying additional resources. This may reduce efficiency, bringing CVI closer to 1.0.

A typical project status spreadsheet as seen in many organizations is shown in Figure 109. Note that all of the previously described information is present in this spreadsheet, in addition to several different assessment methods that are utilized to calculate related variables with regard to project status. These variables include Budget At Complete (BAC), Estimate To Complete (ETC), Estimate At Complete (EAC), Variance At Complete (VAC), and others (See Figure 110.) The interested reader can find much more information available at www.DAU.mil.

	Jan-03	Feb-03	Mar-03	Apr-03	May-03	Jun-03
	$	$	$	$	$	$
BCWS	122,148	263,656	417,065	597,229	811,730	1,043,605
ACWP	138,901	296,993	459,920	654,854	888,258	1,132,596
BCWP	136,064	237,015	373,080	511,339	730,798	959,035
BAC	2,194,588	2,194,588	2,194,588	2,194,588	2,194,588	2,194,588
CV	-2,837	-59,978	-86,840	-143,515	-157,461	-173,561
CV %	-2.09%	-25.31%	-23.28%	-28.07%	-21.55%	-18.10%
CPI	0.980	0.798	0.811	0.781	0.823	0.847
SV	13,916	-26,640	-43,985	-85,890	-80,932	-84,570
SV %	11.39%	-10.10%	-10.55%	-14.38%	-9.97%	-8.10%
SPI	1.114	0.899	0.895	0.856	0.900	0.919
EAC 1	2,240,346	2,749,940	2,705,414	2,810,533	2,667,443	2,591,754
EAC2	2,025,419	3,025,646	2,970,149	3,172,625	2,864,478	2,720,426
VAC 1	-45,758	-555,352	-510,826	-615,945	-472,855	-397,166
VAC 1 %	-2.09%	-25.31%	-23.28%	-28.07%	-21.55%	-18.10%
VAC 2	169,169	-831,058	-775,561	-978,037	-669,890	-525,838
VAC 2 %	7.71%	-37.87%	-35.34%	-44.57%	-30.52%	-23.96%
ETC	1,886,517	2,728,653	2,510,228	2,517,771	1,976,220	1,587,830
ECD	19-Nov-03	5-Feb-04	7-Feb-04	21-Feb-04	5-Feb-04	29-Jan-04

Figure 109: Project Status Spreadsheet

Earned Value Terminology

EVMS	–	Earned Value Management System
WBS	–	Work Breakdown Structure
OBS	–	Organizational Breakdown Structure
Control Account	–	Intersection of WBS and OBS where costs are accumulated and performance measured
Work Package	–	Subdivision of work within a Cost Account Provides level for Earned Value assessment
IMP	–	Integrated Master Plan (Event, Accomplishment, Criteria)
IMS	–	Integrated Master Schedule (IMP + Task)
CP	–	Critical Path: A continuous path of activities in a network having the longest total duration and thus defining the minimum duration of the project. The path with zero float.
Forward Pass	–	Determines EARLY DATES for each activity – the earliest each activity can be done, based on logic and duration
Backward Pass	–	Determines LATE DATES for each activity – the latest each activity can be done without impacting the project end date, based on logic and duration
Total Float	–	The amount of time an activity can be delayed or expanded before it impacts the Project Finish Date
BCWS	–	Budgeted Cost of Work Scheduled
BCWP	–	Budgeted Cost of Work Performed (Earned Value)
ACWP	–	Actual Cost of Work Performed
BAC	–	Budget at Completion
EAC	–	Estimate at Completion (ACWP + ETC)
ETC/CTC	–	Estimate to Completion/Cost to Complete
LRE	–	Latest Revised Estimate (see also EAC)
VAC	–	Variance at Completion (BAC – EAC)

Sample IMP/IMS Structure

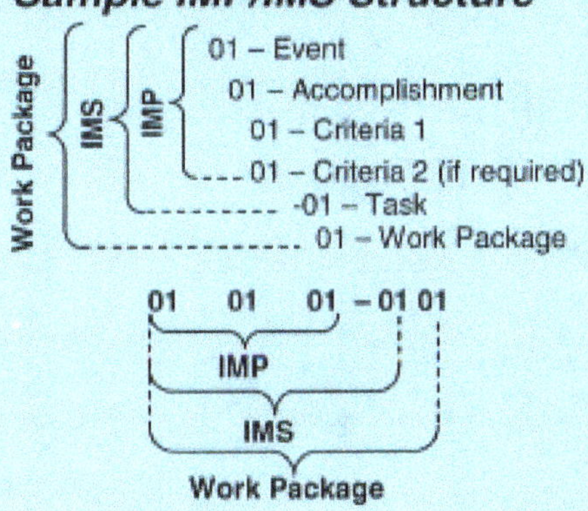

2005

Figure 110: EVMS Terminology

Earned Value Formulas

Cost Variance (CV)

CV = BCWP - ACWP
(60) = 1520 - 1580

$$CV\% = \frac{CV}{BCWP} = \frac{(60)}{1520} = (4\%)$$ 60 units overrun

Schedule Variance (SV)

SV = BCWP - BCWS
30 = 1520 - 1490

$$SV\% = \frac{SV}{BCWS} = \frac{30}{1490} = 2\%$$ 30 units ahead of schedule

Cost Performance Index (Efficiency)

$$CPI = \frac{BCWP}{ACWP}; \frac{1520}{1580} = 0.96$$

Performed 0.96 units for every 1.00 unit incurred.
Index > 1.0 = under cost
Index < 1.0 = over cost

Schedule Performance Index (Efficiency)

$$SPI = \frac{BCWP}{BCWS}; \frac{1520}{1490} = 1.02$$

Approximately 2% ahead of schedule
Index > 1.0 = Ahead of schedule
Index < 1.0 = Behind schedule

To-Complete Performance Index

TCPI is related to CPI and tolerance is within ±5%.

$$TCPI(LRE) = \frac{BAC - BCWP}{LRE - ACWP} = \frac{\text{Working Remaining}}{CTC}$$

TCPI measures the to - go performance efficiency required to meet the declared LRE.

$$TCPI(LRE) = \frac{5000 - 1520}{5100 - 1580} = \frac{3480}{3520} = 0.99$$

Estimate at Completion (EAC) Formulas

1 Weighted Cost/Schedule EAC (80/20) Method

$$EAC = Actuals + \left[\frac{BAC - BCWP_{cum}}{(.8 \times CPI) + (.2 \times SPI)}\right]$$

EAC projection based on past cost and schedule performance.

2 Cum CPI Method

$$EAC = \frac{BAC}{CPI_{cum}}$$

Most optimistic method. Assumes program will perform the same in the future as in the past.

Earned Value Management Data

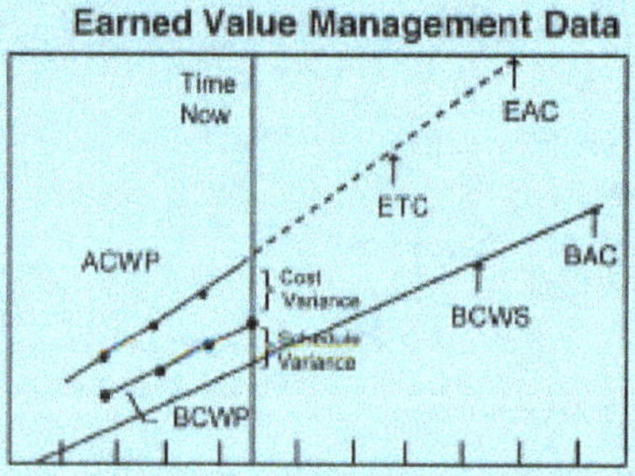

2005

Figure 111: EVMS Formulas

In the event you find a challenge that cannot be met in terms of internal project manipulation, you then resort to Trade-Off Analysis.

Trade-Off Analysis

Successful Project Management is an art and a science. In some cases the Project Manager finds himself in a situation where the time, cost and performance issues are impossible to maintain without project plan modifications. Here a Trade-Off analysis is used.

Six Steps to Project Reconciliation:

1. Recognize and understand the basis for project conflicts

   Human Errors, uncertain problems,

   Unexpected problems
2. Review the project objectives

   From the perspectives of all stakeholders
3. Analyze the project environment and status

   Detailed analysis of planned/actual cost/time/performance (EVA)

   Assessment of potential resolutions
4. Identify possible courses of action

   Brainstorming / Discussions
5. Select the best alternative

   Define best plan of action

   Define impact on Cost/Time/Performance

   Negotiate with Stakeholders for new plan constraints
6. Revise the project plan and publish it to all stakeholders

   Maintain signature approvals as required

Chapter 12: Technology Roadmaps

The development of a Technology Roadmap is a necessity for any Engineering enterprise that seeks to get ahead in the development curve such that the technology developments are readily prepared when the need for inclusion in Product Development has arrived. The methodology has been created that illustrates all of the critical aspects necessary to complete the Technology Roadmap.

- Building a Technology Roadmap
  - Define the constraints
    - What is the current product line?
    - What are the current sales?
    - What are the historical investment levels?
    - What are the human resource limitations?
    - What cultural considerations are there? What appetite for change exists?
  - Voice of the customer
    - Define the current customer base
    - Establish sustaining innovation plans
      - What development is necessary to maintain existing customer base
    - Establish disruptive innovation potential
      - Develop a lower cost solution for over-served customers
      - Develop a different measure of performance for new customers
      - TRIZ as a development method
    - Gaps analysis of existing products / potential markets targeted
  - Establish existing organizational strategy
    - Define if new directions/customers requires shift in strategy
    - Strategy choices
    - Project alignment with organizational strategy
  - Clarify the above to define market segments / products of interest
  - Begin Technology Roadmap definition

Define all users / stakeholders

Define user needs

Establish Functional Analysis for new products

Define services to be delivered for new products

Establish System Architecture for each product

Perform gaps analysis of existing technologies / needed technologies

Map product introduction and lifecycle onto timeline of interest

Map needed introduction dates of needed technologies

Plan technology approaches to develop technologies

Plan projects / resource load to establish development timeline

Map technology development projects onto timeline

The Technology Road-Mapping Process will be further described with examples provided in the ensuing paragraphs. It is the intention of this section of the text to provide Engineering leadership with appropriate methods and processes by which the Technology Roadmap may be created. This construction is based on the premise that a Product Roadmap has previously been generated by the program management function within the organization. If this has not been completed, the current method must be expanded to include targeted customer needs in order to define the products that will be generated from the technologies developed.

First and foremost, in order to begin a Technology Roadmap, an appropriate team must be formed from within the organization. This team should represent a cross-section of the major functions of the organization, and in a typical manufacturing enterprise would include members from the Operations group, Manufacturing personnel, Engineering, Sales personnel, Marketing, and Service, to name a few. These members are selected largely as representatives of their functional areas within the development in order to ensure all facets of the potential products and technologies are considered with regard to the constraints and limitations of the organization.

The next major hurdle in the development of the Roadmap is to effectively define the constraints of the problem or challenge faced by the organization. To accomplish this, we need to examine the current status of the organization in detail and produce a coherent understanding of the existing customer base and products. Therefore we will define:

What is the current product line?
– Define the products and families of products
–Define the basis by which we compete in each of these product areas

What are the current sales?
– Where is the priority of revenue??
– Which product families produce the most return?

An example of this type of development is the advent of the new company known as Soniq Acoustic Systems. Three Engineers from Lockheed Martin Corporation founded the company in 2012. The focus of the effort was to establish a unique and innovative ash removal system to be used in large-scale coal fired power plants. The system was piloted in the Coal Creek plant in North Dakota, which at the time was the largest power plant of its type in North America.

SONIQ
ACOUSTIC SYSTEMS

Figure 112: Soniq Acoustic Systems Pilot Site

In order to fully understand the Systems Development associated with this project, some background information is appropriate.

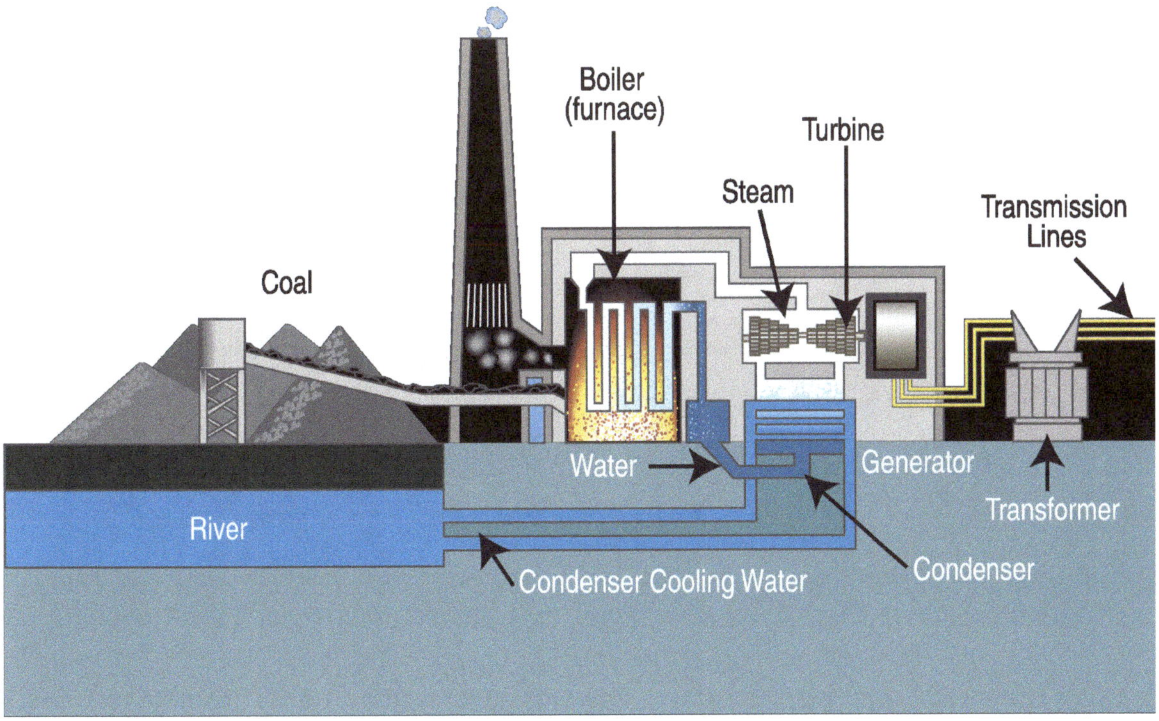

Figure 113: Coal Plant Schematic

In

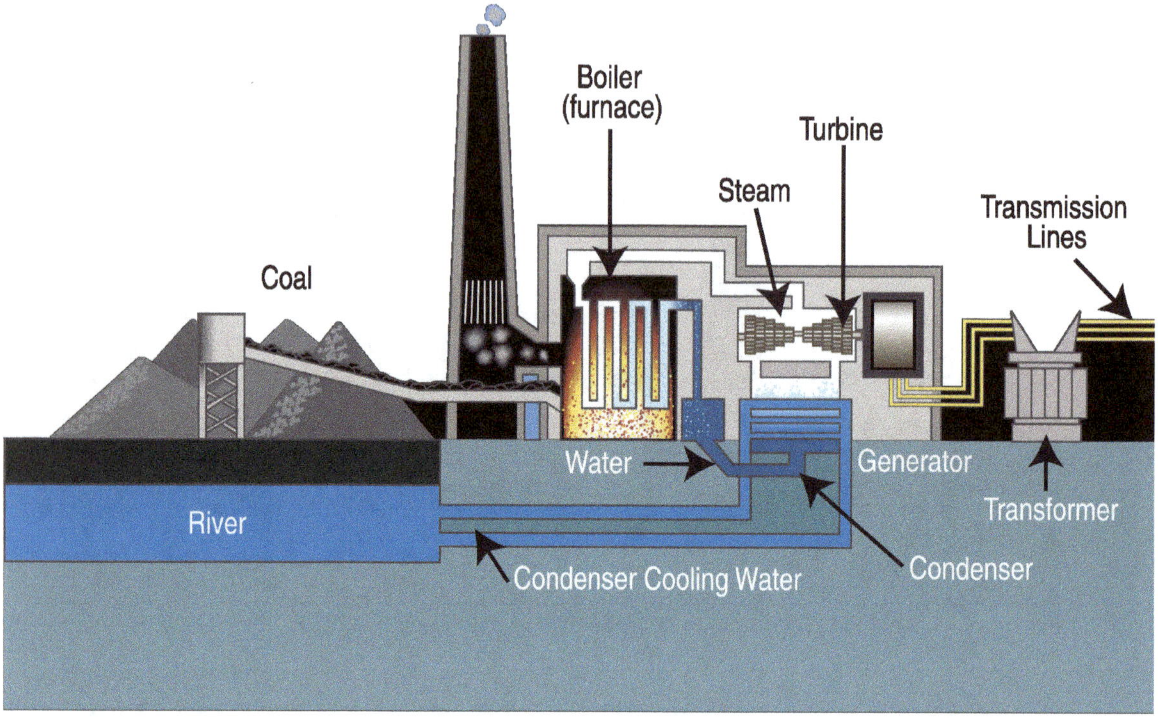

Figure 113, a coal-fired Power plant is shown. Power is generated by means of steam turbines that are in turn powered by large-scale boilers that in turn are fired by aggregate coal, introduced into the center of the boiler and allowed to combust in a gigantic fireball. As the coal is burned, impurities and waste products in the coal are collected on the surface of the water pipes that run through the boiler proper. Over time, ash collects on the inner surface of the boiler cavity, as its surface contains a great deal of declivities and crevices caused by the desire for maximum surface area of the water pipes and mounting of the same. The ash is in a powdery form until the boiler is shut down, at which point the oxygen content in the air combines with the ash to produce a concrete-like substance. In general, a boiler must be cleaned when the thermal efficiency has been reduced due to the ash deposits on the pipes, as the ash acts as an insulator. It is more profitable to shut down and clean the boiler than it is to continue operation at reduced efficiency.

The typical cleaning operation takes a relatively interesting form. In one method that is utilized on a regular basis, associates with shotguns are located at the various ports of the boiler and they proceed to shoot the ash with the shotgun pellets, thereby removing the ash after solidification from the pipes and fittings inside the boiler. Unfortunately, this often causes problems with leakage and damage to the pipes, as well as damage to the associates who will typically fire off an entire pallet load of shotgun shells to clean a single boiler. A second method sometimes used accomplishes the cleaning task with the use of detonation cord. In short, detonation cord is laid throughout the surface of the boiler and exploded, thereby removing the ash deposits from the pipes as well as causing significant damage within the boiler itself.

Soniq was formed in order to accomplish the boiler cleaning prior to the solidification of the ash deposits. This required that the cleaning process be active during the operation of the boiler, which created a large number of challenges for the System Development team. After considerable efforts and research, a system was designed that utilized low-frequency sound waves to cause resonance within the boiler proper, which resulted in the separation of the ash from the boiler walls very effectively. In initial testing, the acoustic system employed resulted in the removal of fourteen tons of ash in the first hour. This was considered a remarkable achievement, and resulted in several contracts for Soniq Acoustic Systems. The development of this system will be used as an example for the Technology Road-Mapping exercise.

We know must establish the constraints of the problem to be addressed by new technology development. We ask ourselves:

What are the historical investment levels?
– Examine the volatility of investment over time
– Predict likely future investment levels

What are the human resource limitations?
– How many new projects can we accomplish?
– How many engineering hours are available for development?

What cultural considerations are there?
– What appetite for change exists?
– Is the organization supportive of disruptions and change?

Once these considerations have been addressed, we have bounded the capability of our Engineering contributions to the challenge. This will provide us some idea of the magnitude of change that can be readily undertaken by the organization, or alternatively, establishes appropriate staffing levels for the change desired.

We must now consider the voice of the customer. This enables us to understand how we are serving our current customers, and begins to establish the product strategy of the organization.

Define our current customer base
– Who are we currently serving?
– What are we providing to them?
– At what frequency are we selling products?

At Soniq, research was focused on the opportunities afforded by large-scale power plants in the United States. This indicated a potential four-hundred-million-dollar market. If the market were expanded into China and India, the potential was easily several billion dollars. The industry's general boiler cleaning process generated a three-day shutdown. At a one million-dollar loss for every hour shut down, the opportunity available in the marketplace seemed considerable.

Given an understanding of the customer and current product line, our position in the marketplace has been defined. We have considered the constraints associated with the problem, and also explored the potential of new marketplaces. Knowing that we have a large opportunity in the identified markets, we now must choose an approach that will result in market capture.

Clayton Christensen has outlined the two preferred approaches to addressing a disruption in a marketplace. If the marketplace is well established, the two choices available to the Company entering the market are to provide:

1) A low-end disruption, which addresses over-served customers with a lower-cost business model
2) A new market disruption, which fundamentally changes the manner in which the customer is served

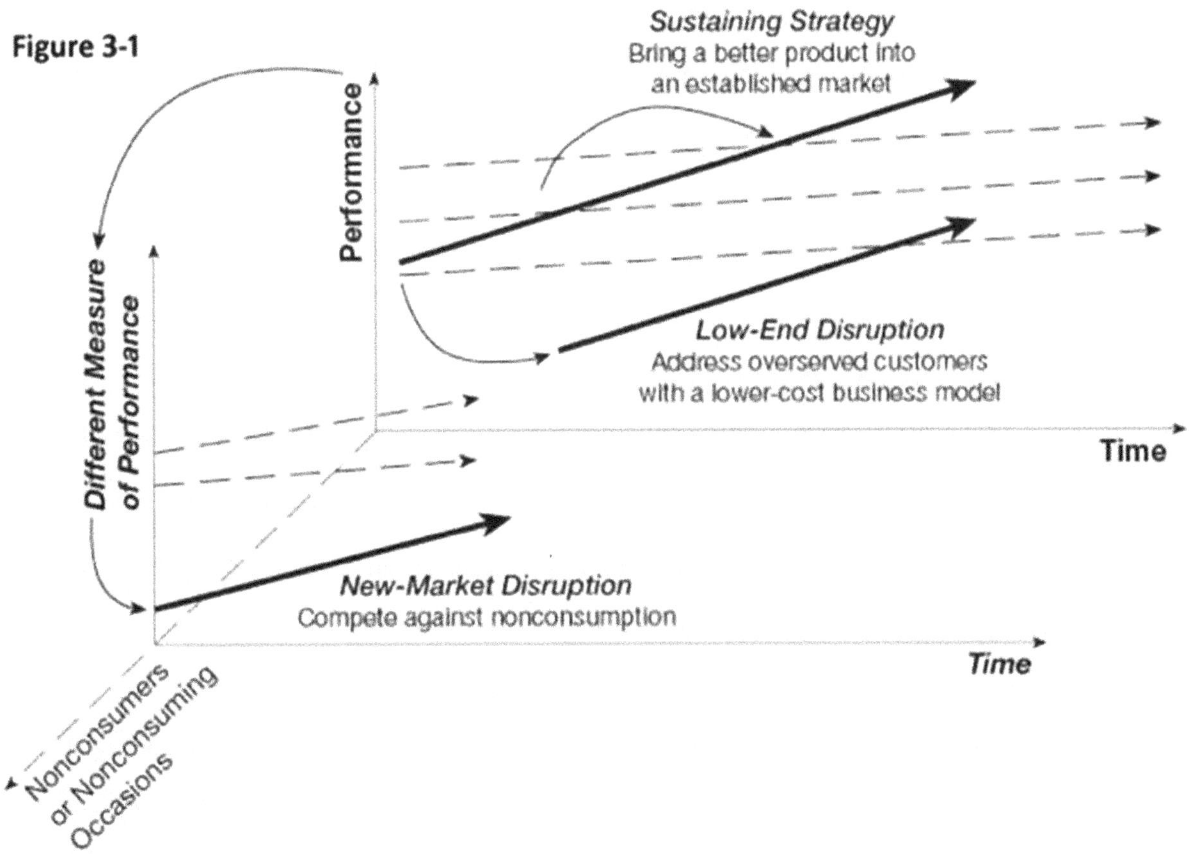

Figure 114: Disruptive Technologies (Christensen)

It is also necessary to maintain the existing product line in parallel with the innovations planned for the organization. Thus we must define and plan those efforts that are necessary to ensure that current customers remain sold on the equipment. Keeping customers sold is a crucial focus for engineering leadership. Once this is complete, we can establish the disruptive innovation potential available to the organization:

Develop a lower-cost solution for over–served customers
Develop a different measure of performance for new customers (change the game)

Dependent upon our organizational strategy, one of these approaches may be preferential to another. Regardless of specific approach, an innovative solution must be defined that will enable the organization to enter the desired marketplace. There are several different approaches available for creative and innovative approaches to problem solving, and one particularly effective method is known as TRIZ.

TRIZ Is a problem – solving, analysis and forecasting tool derived from the study of patterns of invention in the global patent literature. TRIZ was developed by the Soviet inventor and author Genrich Altshuller and his colleagues beginning in 1946. In English, the name is rendered as "The Theory of Inventive Problem-Solving". The system has been used successfully in a wide variety of industries to define disruptive approaches to technology developments. For further information, the reader is directed to the text 'Systematic Innovation' as noted in the references.

Next, we conduct any gaps analysis of our existing products and the potential markets that have been targeted. This is accomplished after consideration of the disruptive approaches that might define potential solution areas or regions of research to be pursued. Coupled with our understanding of the current system limitations in our marketplace, we should be able to identify various gaps in our offering that might give us additional traction or a unique solution to an existing shortcoming or problem in the marketplace.

In the Soniq example, we understood that industry cleaning technologies required a boiler shutdown, were typically conducted in a dangerous manner and resulted in a significant productivity loss for the organization. From this we can establish several requirements which will enable a disruptive approach:

Provide a system that operates with the boiler active
Make the system automated, allowing regular periodic cleaning on a cycle considerably shorter than the current biannual approach
Do so in a noncontact manner, staying out of the boiler for safety reasons

With an understanding of the gaps we intend to rectify, we seek to find a more appropriate strategic approach. For more clarification on defining the strategic focus for a project, see the earlier chapter on that subject. In the case of Soniq, the organization elected to provide a unique and innovative solution that would enable higher production efficiencies, increased safety and an opportunity for the organization. Therefore innovation was the selected strategy. The organization needed to bring the coal plant a solution that no one else could offer, and by virtue of that, charge fees accordingly.

Once we have defined our strategic interests, we need to clarify the appropriate market segments and products of interest for our development efforts. This implies that we need a good understanding of where the gaps are in the current market and a good understanding of how we might shift the playing field. From that mindset, we define our needed product offerings.

In the Soniq example, the organization elected to develop a system that would utilize low-frequency sonic energy to introduce vibrations into the boiler. If this could resonate the pipes or the ash, it could cause them to separate, and the ash to drop into the collection hopper. The amount of energy required was established based upon the fact that the system needed to accomplish this task from outside the boiler and while the boiler was in operation. This defined the technological approach to be utilized to produce the system capabilities necessary to remove ash from an active boiler system.

At this point we are ready to begin our Technology Roadmap definition. To accomplish this we will use a Systems Engineering approach to the Roadmap definition, as it provides a multiple viewpoint aspect understanding the system to be developed. This adheres to INCOSE standards and has been proven to avoid costly and repeated incremental investments and instead enables a longer-term perspective in Technology Development.

We must first define all of the users and stakeholders for our system. To do this we consider all of the individuals and groups that are impacted by our program or project. We then establish their interests in the effort and thereby define what capabilities need to be delivered to satisfy their needs.

Based upon this list of capabilities to be delivered, we utilize typical Systems Engineering approaches to define a Functional Analysis and complete decomposition of the system in an effort to produce a viable System Architecture. From the System Architecture, we can begin to develop our physical architecture, enabling a tangible manifestation of the capabilities requested. In the process of developing the Physical Architecture, we will perform a Gaps Analysis of existing technologies as well as needed technologies. This is accomplished by examining the functional capabilities required in our System Architecture and defining what technologies already exist for purchase and/or acquisition. We then define areas where the technologies are not readily available, and conduct research to understand where they are in their development cycles. We can also establish the probable risk for development of those same technologies within our organization, which will include the magnitude of investment as well as the probability of success. This effort will establish our make/buy decision for technologies needed.

At the same time we must consider the hierarchy of technology development. This includes a five-stage process as defined in the Systems Engineering section of this text. Those five levels include:

Concept Refinement stage
– Development of new science and/or concepts

Technology Development stage
– Establish if the fear radical system will function effectively

System Development and Demonstration stage
– Complete the technology development and prototype, Ultimately getting the system to release

Production and Deployment stage
– Small lot production and refinement

Operations and Support stage
– Logistics and sustainment for products

We then map the product introduction and product lifecycle onto a timeline of interest. Typically we would define the duration of our Technology Roadmap to be something on the order of three to five years. We then define the elements of the product roadmap that are pertinent to our plan. If the products for the Product Roadmap are not available, then we must define the products that will be introduced (as an expansion of our Roadmap), when they will be offered, and what the product lifecycle is expected to be. The beginning of a Technology Roadmap for Soniq Acoustic Systems is shown in Figure 115.

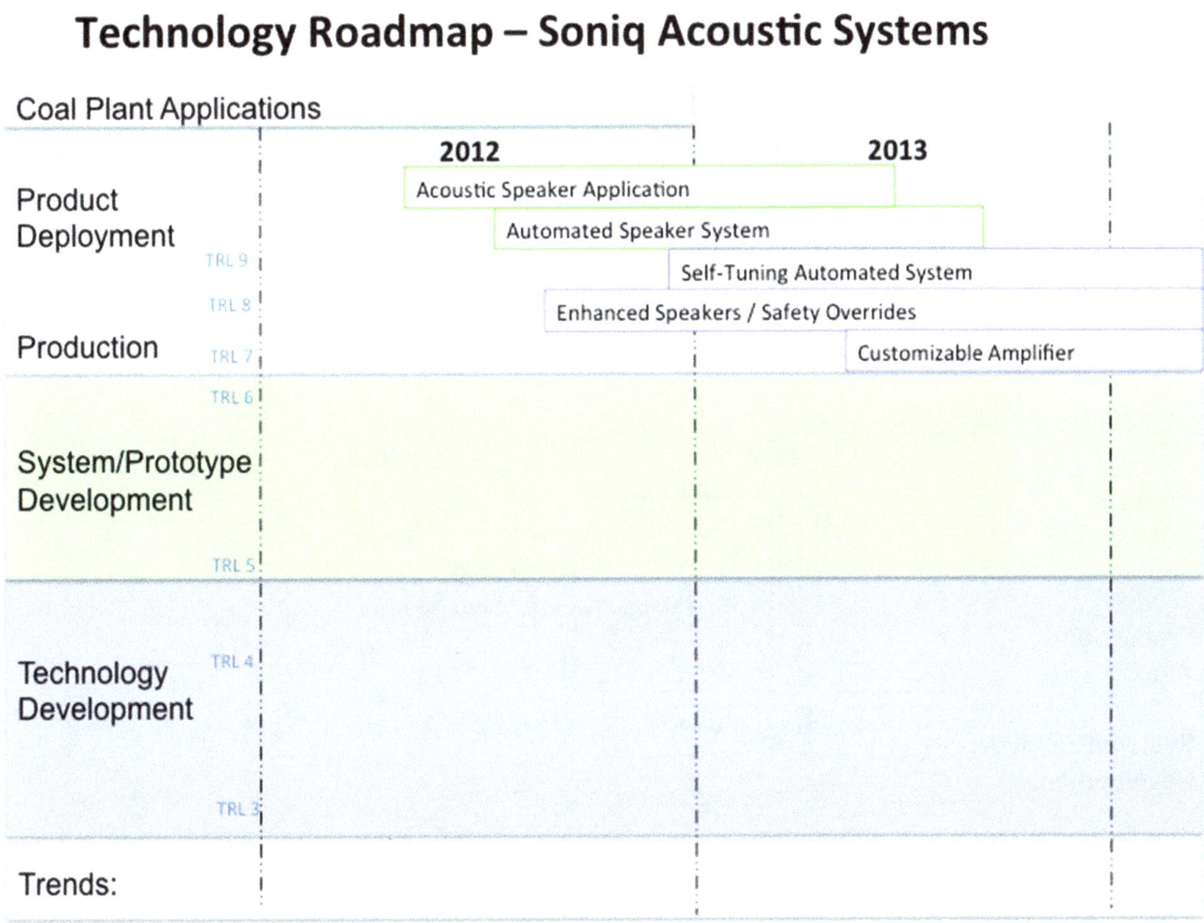

Figure 115: Product Focus

We then map the needed introduction dates of the technologies required to support these products. In order to enable the desired products, we must define the necessary technologies needed to bridge the gap. A new form of that technology must be delivered in order to deliver the product capability. We will also add in any existing development projects that are required to complete the required System Developments. Finally, we must define the system projects to be accomplished to complete the Product Development.

The system projects must then be planned in detail in order to develop the necessary technologies. To begin, we establish the scope necessary to accomplish the purpose for our products. We then define our project plans, including the specification, schedule, and budget. We then establish the risk factors for the project completion, and refine the plans as necessary to establish a reasonable probability of success.

It is then necessary to resource load the System Development plans in order to establish a development timeline. Thus, we must consider the multiple projects being accomplished simultaneously, as well as the Human Resource limitations of our Engineering organization. In parallel with this, we must consider the budgetary constraints imposed by the organization. At this point we can place the projects on the timeline to create a practical total workload for the development group. Last, we should establish any apparent trends that may exist in the development cycle. This is shown in Figure 116.

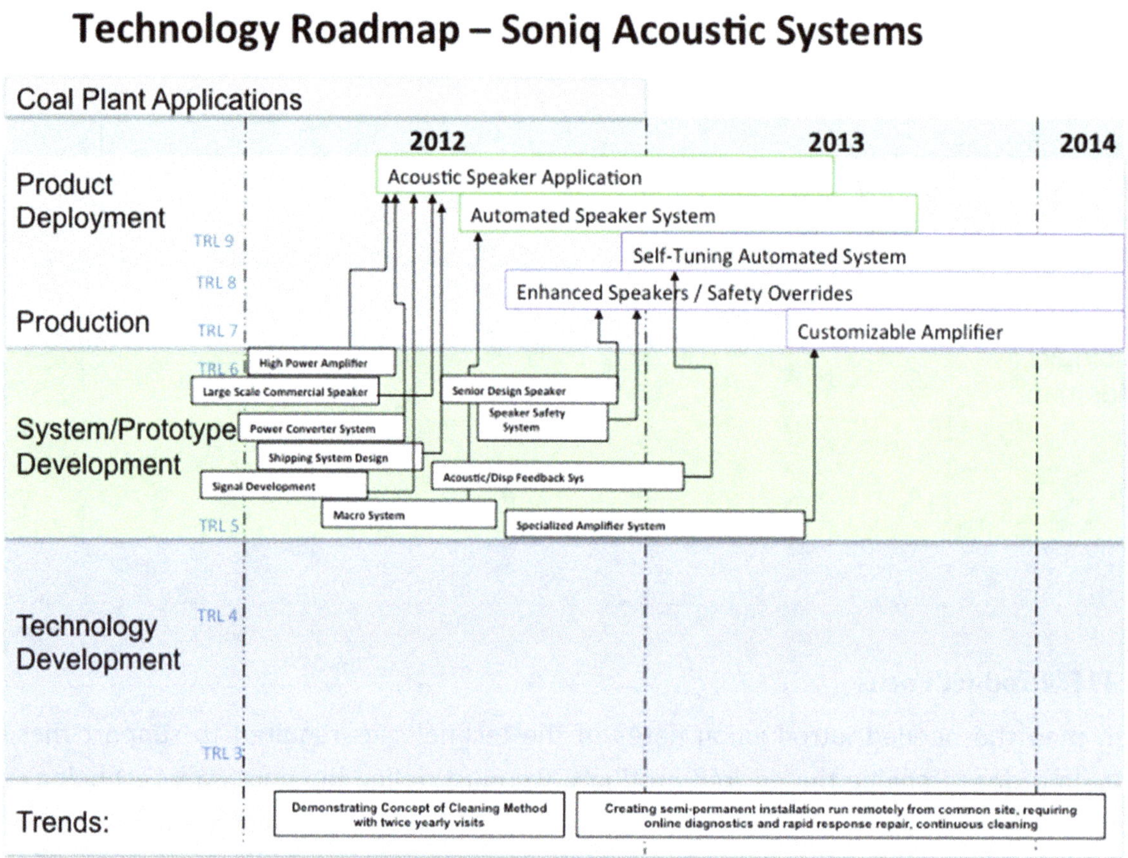

Figure 116: System Developments and Trends

We then complete the same approach for the technology development level. We establish the needed technologies, and establish when they are needed in order to accomplish the project developments. We then plan the technology development projects including the specification, schedule and budget necessary, and finally established the risk plan for the technologies to be developed.

The technology development projects must be mapped onto the appropriate timeline. This is accomplished by doing the resource loading for the necessary technology development projects, and load-leveling the multiple projects that are active at any given time. One item of note is that this may require prioritization and trade-off analysis based upon the workload introduced into the system. Finally, we define when each project must begin in order to provide appropriate delivery dates for the projects. This final step is shown in Figure 117. As can be seen, this produces a chronological map of the technologies needed to be developed, which then inject capabilities at the system level, which in turn enable the introduction of products to the marketplace. A Roadmap of this type is very useful to define for many in the organization that major advances do not happen overnight in the Engineering department.

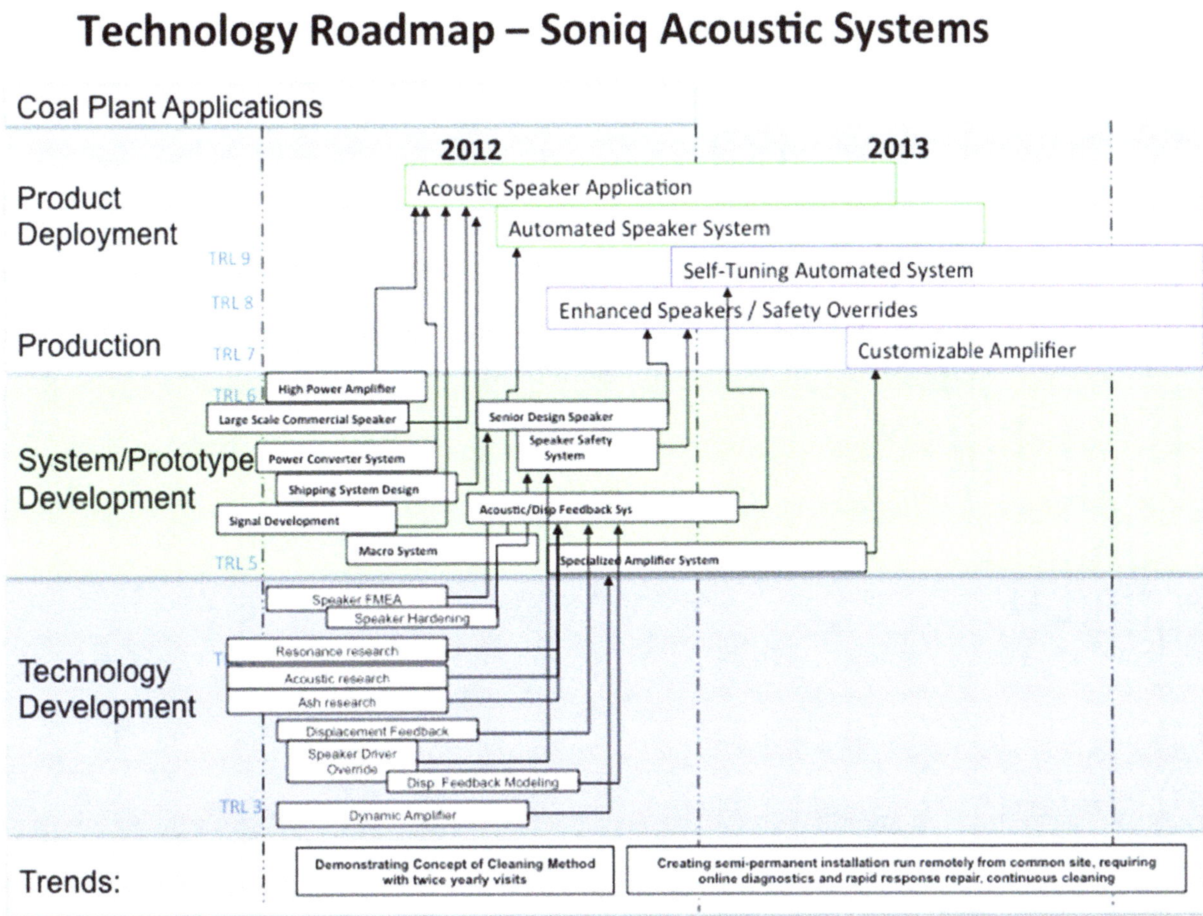

Figure 117: Technology Roadmap for Soniq Acoustic Systems

Once the plan has been completed, we undergo a series of reviews for Roadmap and Development plan refinement. These should include individual meetings with the stakeholders of the Technology Roadmap, including Operations, Manufacturing, Sales, Marketing, Engineering and all other functions. These meetings are utilized to assess and revise the plan as appropriate in order to finalize our Technology Roadmap.

Finally, we must schedule the development of the Technology Roadmap much in the same way as recommended in the Project Management section of this text. A proposed Technology Roadmap development cycle is shown in Figure 118.

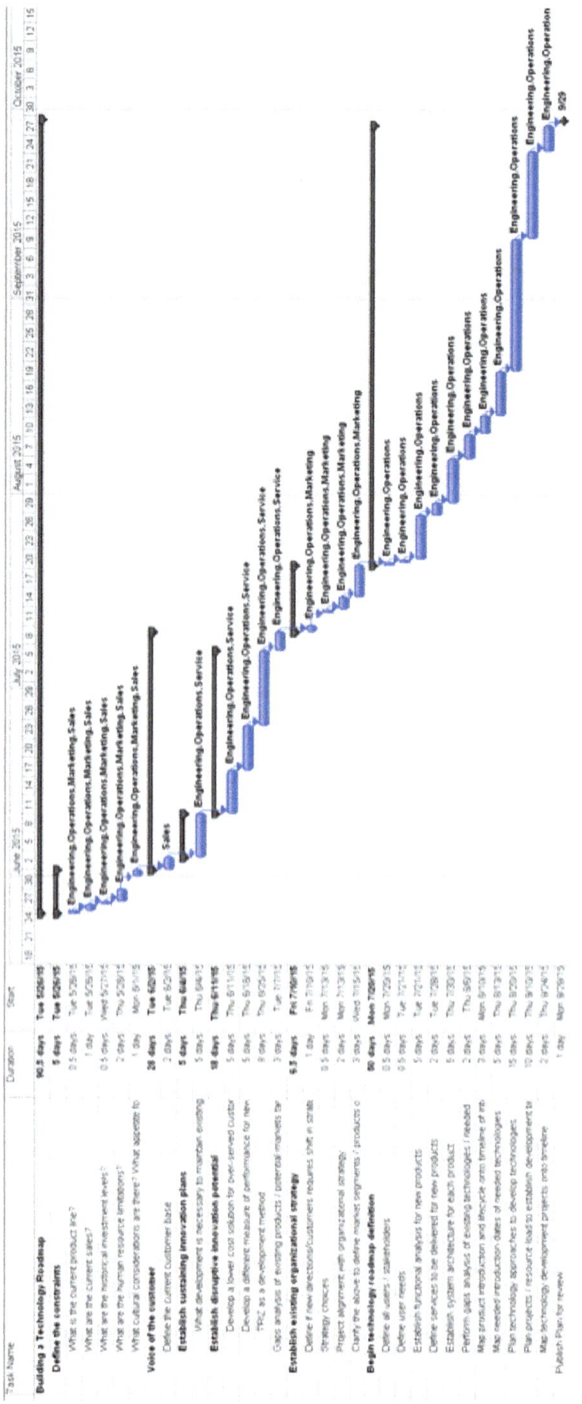

Figure 118: Technology Roadmap Definition Plan

Chapter 13: Personnel Selection

One of the most crucial aspects of Engineering leadership is the ability to select and hire the most effective talent possible. This includes finding people that have exceptional technical skills or capabilities (perhaps not available from other associates in the organization), but may also include many intangible or less obvious factors that must be taken into consideration.

These factors will include things like exceptional communication skills, writing skills, presentation preparation and delivery, and interdepartmental effectiveness. Some of these factors are relatively easy to assess in an interview, while others are virtually impossible to assess practically before having significant opportunity to observe the performance of the associate.

As might be expected, we have spent considerable time during the performance appraisal process assessing and developing a description of the behaviors that are most desired in a given position. We will of course use those conclusions and results as a significant input into the personnel selection process. These behavioral elements would enable us to assess the candidates expected job fit within the organization. Additionally, we will develop a description of our organization in order to establish a cultural description of the department, which will allow us ultimately to assess the cultural fit of the candidate, as well.

KNOWING HOW PEOPLE FIT THE ORGANIZATION

We must consider what effect culture has on the organization. Culture is really nothing more than the common descriptions people have about the organization:

- The common behaviors
- The common actions
- The way it is done around here
- What makes us different than our competitor

Culture is made up of a variety of elements, but can be considered to be similar to the personality traits of an individual. By virtue of that several models exist that enable us to assess the type of culture and strength of culture within our organizations.

As an example, we can consider Robbins' perspective on culture, which is made up of:

Innovation and Risk Taking

How much the individuals within the organization are encouraged to take risks

Attention to Detail

How much precision or rigor is required in the day-to-day efforts of the organization

Outcome Orientation

Whether the individuals are results rather than process driven

People Orientation

How management typically makes decisions with respect to people; is it all about the bottom line, or do people matter?

Team Orientation

How we are organized with respect to teams

Aggressiveness

Whether individuals in the organization to focus upon competition or agreeableness

Stability

Whether we are comfortable with the status quo or focused upon growth

This can also be considered to be a measure of how much Chaos we're comfortable with

We need to define a practical method that will enable us to choose the most effective individuals for our organizations. This method must take into consideration the organizational structure, individual behaviors, and organizational culture and job fit in order to effectively assess whether or not the individual is likely to have high job performance. We have already considered organizational structure as well as job performance behaviors, and will therefore focus on cultural assessment and fit.

To accomplish this, we will examine the factors defined by Robbins and assess our organization as being neutral, higher or lower in that particular focus area. This will enable us to create a diagram similar to that shown in Figure 119, which will clarify for us what items of importance we may choose to explore with regard to our interviewing process. This assessment should be based upon evidence and past performance in order to a sure that our assessment is in alignment with current cultural behaviors.

For example, in this case we may choose to identify a strong outcome orientation and a strong attention to detail as being the most critical factors for the success of our organization. We may also choose to note that risk taking is not a strong element of our culture.

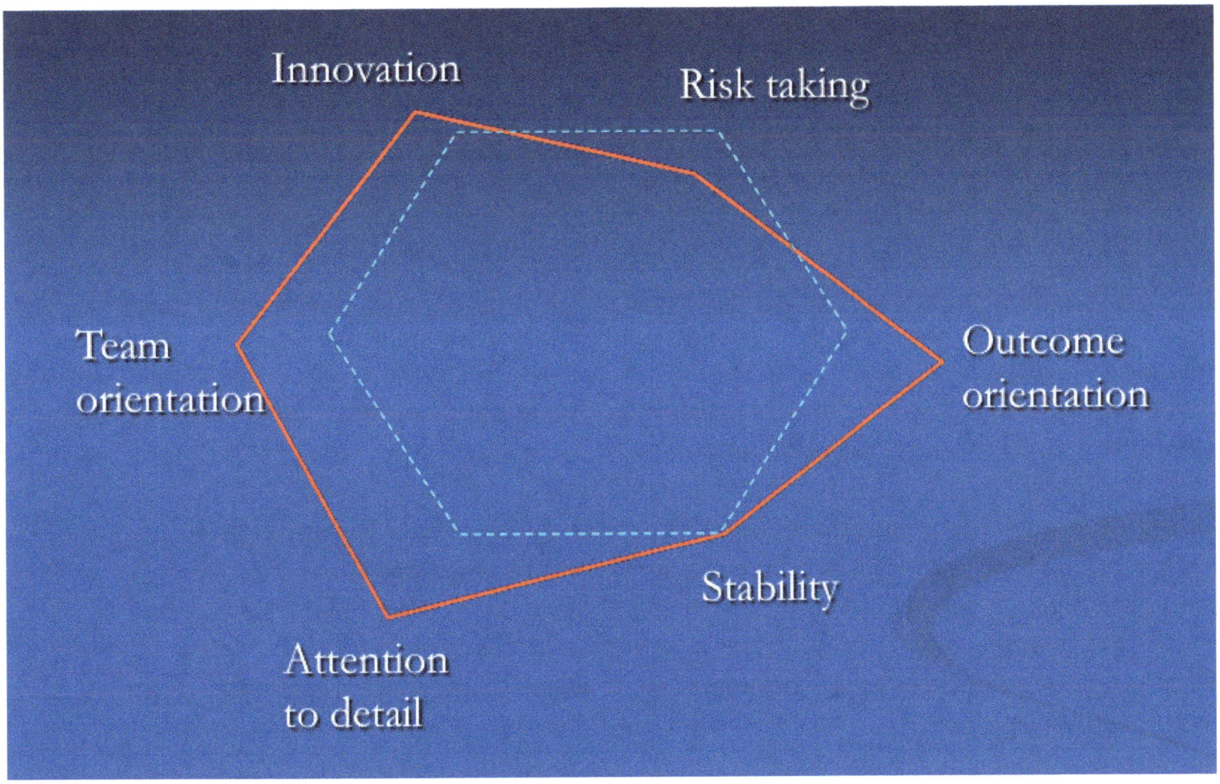

Figure 119: Organizational Culture Assessment

After an assessment of our organizational culture, we can convene a group of individuals who will be focused on creating an appropriate hiring strategy for personnel to be added to the engineering organization.

Culture defines boundaries, provides a sense of identity, and fosters commitment to the whole of the organization. It also increases social system stability and provides sound sense making for individuals unclear on the causal factors for actions within the organization.

A crucial factor in assessing the quality of a culture within an organization is the determination of whether or not the strategy of the organization is clearly present in its culture. Culture can also be a liability to an organization, as it can create a lack of movement or initiative. It can stifle creativity and may drive out innovators. In a case such as this, the company may become entrenched in the manner in which it does things, and can lose its long-term competitiveness.

Culture is passed on within the organization through the process of socialization. This can occur in many different ways over a wide range of timeframes. By virtue of this, changing an organization's culture can be a very time-consuming and challenging task. In general, cultural change is typically best introduced when the organization is already undergoing a disruption for one reason or another.

Changing the stories or rituals that are utilized in its socialization process for new employees may change an organization's culture. One of the simplest ways to accomplish this is to change the acronyms, jargon and language used within the organization, which will result in people needing to change their perspective in order to continue communicating in an effective manner.

Given the previous example of defining the culture necessary to accomplish a candidate selection from amongst a group of qualified candidates, we need to complete our personnel selection criteria and methods in order to more effectively choose employees for the organization. Thus, we find that we must follow a systematic method in order to explore all the elements of interest in the selection process:

- Define what is important with respect to the behaviors of the individual
- Define how to measure those behaviors – Performance metrics
- Define our organizational structure and identify the aspects of interest
- Clarify our culture and identify the aspects of interest
- Construct a tool to explore the behaviors, organizational and cultural issues of the candidate

In the end, we intend to create a detailed questionnaire that will be utilized by the interview team when conducting interviews for various positions within the Engineering organization. This detailed questionnaire should address the aspects of interest that we choose to explore related to our organization, culture and expected behaviors. In general, we will interview based upon a structured process. This implies that all of the interview questions are scripted and prepared beforehand in order to assure that all candidates can be compared after completion of the interviewing process. This will increase the validity our conclusions by a significant factor, and will remove some of the personal bias many hiring managers feel towards candidates. It will also cause us to reflect upon our organization's characteristics, which may enable us to find ways to incrementally improve the organization.

HIRING THE RIGHT PERSONNEL, AND WHY

It is very important for us to recognize that the proper selection of personnel requires many of the skills necessary for a manager to effectively manage his or her organization. This is due to the fact that proper personnel selection requires the manager to understand the jobs she supervises, as well as the behaviors appropriate for those jobs to be effectively accomplished. If a manager spends considerable time defining what behaviors are appropriate for a given position within the organization, he is much better suited to manage the personnel fulfilling that position.

The most common complaint of most new managers is the inability to manage personnel based upon the conception that they have poor attitudes or are incapable of accomplishing their tasks effectively. Thus, the manager might find himself in the position of endlessly trying to make up for the shortcomings of his subordinates.

In reality, many of the shortcomings of subordinates are caused by the manager's unwillingness or inability to provide appropriate feedback to the employees. For example, consider the employee who has demonstrated mediocre performance and has great room for improvement. The manager may find herself in a situation where she does not want to present a negative outlook to the employee, and therefore tends to gloss over the issues of importance in performance shortcomings. When this occurs, the employee is given the impression that his behavior is entirely appropriate for the organization. This results in his continuing to behave in exactly the same manner, or even pushing the envelope to some degree, and expecting further positive feedback.

Before long this unwelcome behavior will transcend the single employee and be exhibited in a variety of associates. Other individuals within the organization will quickly recognize that the manager considers the behavior observed in their colleague acceptable. This is further supported by the broad-brushstrokes that are frequently used in completing personnel evaluations and salary increases amongst many organizations. If all members of the organization receive the same increase in salary regardless of their actual job performance, it doesn't take long for employees to figure out that a considerably less notable performance will result in an equivalent salary increase.

Therefore, if a manager is willing to make the hard decisions and to follow through with the difficult communications of negative feedback to the employees, he will find the job is greatly simplified and the employees will recognize that appropriate behavior is awarded appropriately by the organization. This will result in a work force that is striving to accomplish more, rather than the more common case of individuals attempting to accomplish as little as possible while achieving salary increases that are not commensurate with their levels of performance. Additionally, associates will begin to see the manager and organization as 'fair'. This is crucial to build trust within the team.

An example interview guide that has been created utilizing these considerations is provided for reference.

Engineering Interview Guide ____

Candidate Name Date

This document is intended to provide a framework for the interviewer to produce a structured and coherent interview for potential candidates in the workplace. The vision and values of the organization are used as a guide to shaping our interview approach and methods.

The methodology recommended for interview questions is based upon the STAR model of interviewing. The STAR (Situation/Task, Action, Result) format is used by interviewers to gather all the relevant information about a specific capability required by the job. This interview format is said to have a higher degree of predictability of future on-the-job performance than the traditional interview.

> **Situation / Task**: The interviewer wants you to present a recent challenge and situation in which you found yourself. What did you have to achieve?
>
> **Action:** What did you do? The interviewer will be looking for information on what you did, why you did it and what alternatives were possible.
>
> **Results:** What were the outcomes of your actions? What did you achieve through your actions and did you meet your objectives? What did you learn from this experience, and have you applied this knowledge since then?

Given the Vision and Values in Engineering, we can focus on several areas of exploration in our interviewing process. These include:

<u>Technical Competence / Excellence:</u> Does the candidate possess the knowledge and background or experience to be able to effectively execute the job? Is the level of technical competence appropriate for the position? Does the candidate desire to be an excellent Engineer / Technician?

<u>Drive Solutions for Customers:</u> Does the candidate demonstrate a capability of producing innovative solutions and products? Does the candidate have direct or indirect experience in shaping a future state? Does the candidate demonstrate an understanding of how to provide technical leadership? Does the candidate demonstrate an ability to investigate, formulate and develop practical solutions?

Team Engagement: Does the candidate demonstrate the ability to form an effective team to accomplish complex objectives? Does the candidate understand the needs and focus of the individual team members and can he or she adjust his or her approach to achieve the best results?

Deliver Results: Can the individual resolve challenges and difficulties in order to deliver on plan? Does the candidate have past experiences that demonstrate and ability to overcome adversity? Can the candidate execute projects and plans effectively?

Demonstrate Professionalism, the Park Way: Does the candidate demonstrate a high level of personal integrity and professionalism? Does the candidate focus on building effective relationships and collaborative work environments?

Technical Competence/Excellence:

Describe a recent technical challenge you faced, and its resolution.

Given a customer problem, what approach would you take to achieve your objectives?

Describe a situation where you were required to analyze information in order to identify a need.

Give me an example of when you discovered a more efficient way to perform your job.

Why did you to become an Engineer/Technician, and what are your long-term goals in the profession?

Drive Solutions for Customers:

Can you share with me a difficult problem you had to sort out in your last job?

Would you please describe a new idea or suggestion that you made to your supervisor recently?

Can you tell me about a time that you had to use your judgment to make a tough decision in your previous position?

Walk me through how you make a grilled cheese sandwich.

It can be difficult to get people to think "outside the box". Tell me about a time when you were able to get others to go beyond conventional thinking. What did you say and do? What happened?

Team Engagement:

Can you tell me about a time you persuaded others to accept your idea?

Describe a situation when you were able to successfully influence another person.

Tell me about an unpopular decision you've made, and its result.

Tell me about a time you had a conflict and how you resolved it.

Tell me about a time when you had to work with a group of people who came from different backgrounds. How did you make it work?

Can you describe a time that you demonstrated the ability to be an effective team member?

Deliver Results:

Tell me about a time you had to quickly adjust your work priorities to meet changing demands.

Describe a time when you were faced with a stressful situation that demonstrated your coping skills.

Tell me about a time that you were not satisfied with your work performance or product. What did you do about it?

Tell me about an instance when a Customer made an unreasonable request. What was the request, what did you say or do, and what was the result?

Can you tell me about a situation where you took a leading position?

Tell me about the work you are most proud of, and why.

Can you share the biggest disaster you have had working projects?

Demonstrate Professionalism:

You attend a meeting where another associate raises his or her voice and becomes quite angry toward you. How do you react to this situation?

Tell me about a time you witnessed a fellow employee do something that you didn't think was appropriate. What did you do about the situation?

Tell me about a time when you received negative feedback from your manager or supervisor. How did you handle this?

If you were responsible for a project that was running behind schedule and over budget, how would you approach your team to get improved performance?

The effective Engineering leader will recognize the importance of understanding all the aspects of personnel selection in order to better direct his direct and indirect employees in accomplishing their projects. This includes developing skills in the area of:

- Understanding and evaluating the appropriate tasks required of a given job
- Assessing appropriate behaviors that will aid in accomplishing the tasks required of a given job
- Defining metrics that will be utilized to evaluate those behaviors
- Communicating those metrics to the employee in the form of objectives and behaviors
- Continually assessing the employee's performance with respect to those objectives and behaviors
- Recording performance on a regular basis
- Communicating personal performance effectively to the employee on a continuous basis

Once these tasks have been accomplished, it will be clear to the Engineering leader that one of the greatest favors she may do for herself is to select appropriate people for the organization or project. This will include:

- Knowing what organizational structure is appropriate to accomplish the work
- Understanding the requirements of the job
- Understanding of behaviors necessary to accomplish those requirements
- Defining the existing organizational culture
- Creating a tool that can be used to assess all pertinent factors prior to selecting a candidate

This will enable the manager to greatly improve the reliability and validity of his or her selection process and will greatly improve the quality of the workforce provided for the organization. It will also create a much easier job for the manager, thereby increasing job satisfaction and his or her ability to focus on his or her own job performance as opposed to spending all of his or her time correcting subordinates' mistakes.

Chapter 14: Training and Education

Once we have selected talented candidates for inclusion in our Engineering organization, it is important for us to develop and keep those employees. One manner in which this is accomplished is by creating a dynamic training and education program. There is a distinction between training and education, in that training is typically focused on job-oriented skills presented in a short-term training format, while education is more focused on the attainment of a degree or certificate in order to establish a wider breadth of skills and capabilities, and usually requires a significantly longer development process.

There are many resources available for training. When developing a training program, there are two areas to consider: what to train people on, and how to accomplish their training. Each of these are of significant importance, and the Engineering leader must spend some time in developing appropriate training areas and methods that will be effective for his or her work team.

Understanding what skills to train is usually the first consideration. In principle, it is relatively straightforward: train your individuals of all of the gaps between their current skills and the skills needed to accomplish their jobs effectively. For example, if we have hired an entry-level Engineer who possesses excellent technical skills but lacks Project Management skills, and many other associates lack Project Management skills, we may choose to bring in a Project Management training program..

In practical terms, you simply cannot train everyone for every skill. An associate will need to learn some things on her own, certain skills will be less important to some associates, and there will also be some unanticipated new development areas that develop after associates' training plans are formulated.

The methods selected for training less prescribed. There are many different approaches, including classes, books, online tutorials, mentoring, or throwing the associate into the deep end of the pool and hoping he doesn't drown. In reality, it is probably best to align the training methods selected with those that provide the greatest learning opportunity for the associate. This may be an impractical requirement, however, as most training programs focus on multiple individuals simultaneously, thereby limiting their ability to cater to individual differences. For most Engineering organizations several considerations will constrain the training delivery options:

Budget–

Often the expense of providing formalized training can be hundreds to thousands of dollars for a single day of training for a single employee. While the results might be effective, it is very difficult to justify these costs with a large department of associates that need individualized training.

Timing –

Some training methods are available at any time, such as books and online training. Some training is only offered at specific times or locales. For example, Project Management Certification is offered at a wide variety of universities, but students must take a sequenced set of courses and even a single missed course offering may delay the completion of the certificate by as much as a year.

Availability –

Many skills-specific training areas are very difficult to find available in the public sector. For example, very few training courses are offered in the areas of Systems Engineering or Program Management. Should your associate need development in the skill areas, the Engineering leader will need to be much more creative in identifying appropriate options, such as finding an expert in the field who is willing to mentor an associate.

Focus–

Many training options have been designed for the general population, and may be decidedly less useful for Technical professionals. Often, Engineers complain of entering a classroom only to be told that the instructor has elected to remove the Math from the program in order to not threaten individuals who are less comfortable with the subject matter. Engineers have a need to develop their skills, but unfocused offerings often lead to low enrollment.

In general, Engineers are the less likely to take advantage of training offered by the organization. Although it may be available to ninety percent of the population, it is unusual for more than fifty percent of the Engineers to take advantage of the opportunity. When the training is irrelevant or poorly presented, many Engineers feel justified in this lack of interest in career development. It is incumbent upon the Engineering leadership, then, to address this problem and provide opportunities for the associates in the Engineering department to further their education.

When skill development is required that needs significantly more depth than is offered in a typical training program, it is often most appropriate to pursue a degree program. Many colleges and universities offer degree programs that are topical and useful for an Engineering department. However, it is a significant challenge to partner with a local university or college in order to tailor a development program appropriate for your Engineering organization. For example, in order to develop specific Systems Engineering skills in a manufacturing organization, the organization contacted two local universities and proposed a new degree program in the discipline of Systems Engineering. This effort took considerable time, but in the end, a program tailored to the needs of the organization and many related organizations in the region was successfully implemented. The resultant curriculum for that degree program is included for reference.

System Engineering Program

Program Purpose:

The Systems Engineering curriculum was originally requested by a number of industrial partners in the Twin Cities and outlying regions in order to serve their employment needs. These partners included Industrial, Manufacturing, Medical Device and Defense partners. Collectively, they saw a need to provide educational opportunities at a graduate level for Engineering graduates who had progressed in their careers to the point that they needed to fulfill new roles. These included:

Program Manager
Systems Architect
Systems and Software Engineer
Systems Engineer
Engineering Project Manager
Medical Device Systems Engineer

Program Need:

The Systems Engineering degree is focused on a core curriculum that enables the various roles to be fulfilled by graduates, providing them practical education and experience in the discipline of Systems Engineering. Due to the highly experiential nature of the Systems Engineering domain, special attention was focused on assuring the graduates of the program fulfilled the needs outlined by the collective partners. These include:

A knowledge of and experience with the Systems Engineering process
Project Management skills

Program Management skills
System Architecture knowledge and methodologies
The ability to address highly complex problems systematically
The ability to plan, manage and execute highly nebulous and varying objectives for organizations

Master of Science in Systems Engineering – University of St. Thomas Saint Paul, Minnesota M.S.Sy.E. Degree – 30 Credits required

Core Courses (All Required) – 15 credits

ETLS 507 Introduction to Systems Engineering
ETLS 508 Systems Design
ETLS 509 Verification and Validation
ETLS 601 Program/Project/Team Management
ETLS 789 Modeling and Simulation

Elective Credits (4 required) – 12 credits

(Note that no more than 3 electives can be taken from any concentration area without prior advisor approval)

Financial and Accounting Electives

ETLS 505 Managerial Accounting and Performance Management
ETLS 661 Eng Economic Analysis & Prod Control

Human Aspects Electives

ETLS 670 Masterful Leaders and Leadership
ETLS 671 Human Aspects of Technical Management
ETLS 855 Implementing Innovation

Modeling and Simulation Electives

ETLS 777 Finite Element Analysis
ETLS 785 Simulation Logistics
ETLS 810 Advanced Controls

Software Electives

SEIS 601 Foundations of Software Development
SEIS 610 Software Engineering
SEIS 635 Software Analysis and Design
SEIS 640 Operating Systems Design
SEIS 645 Computer and Network Communications

Medical Device Electives

ETLS 520 Design & Manufacturing in Medical Device Industry
ETLS 722 FDA Medical Device Quality Systems
ETLS 724 Medical Device Clinical Studies
ETLS 731 Combination Products, Drugs and Biologics
ETLS 737 Int'l Regulatory Affairs for Medical Devices

Technical Electives

ETLS 501 Production Operating Systems
ETLS 506 Statistical Methods for Manufacturing Quality
ETLS 672 Excellence in Design
ETLS 701 Design of Experiments
ETLS 744 Power Systems

Engineering Project – 3 Credits

(Must be pre-approved by Advisor)

ETLS 880 Engineering Project (3 credits) Project/Written Paper and Presentation

---

Suggested Courses of study are provided for several common Systems Engineering roles:

Program Manager

ETLS 507 Introduction to Systems Engineering
ETLS 508 Systems Design
ETLS 509 Verification and Validation
ETLS 601 Program/Project/Team Management
ETLS 789 Modeling and Simulation

ETLS 505 Managerial Accounting and Performance Management
ETLS 661 Eng Economic Analysis & Prod Control
ETLS 670 Masterful Leaders and Leadership
SEIS 601 Foundations of Software Development
ETLS 880 Systems Engineering Project

Systems Architect

ETLS 507 Introduction to Systems Engineering
ETLS 508 Systems Design
ETLS 509 Verification and Validation
ETLS 601 Program/Project/Team Management
ETLS 789 Modeling and Simulation

ETLS 777 Finite Element Analysis
ETLS 785 Simulation Logistics
SEIS 635 Software Analysis and Design
ETLS 701 Design of Experiments
ETLS 880 Systems Engineering Project

Systems and Software Engineer

ETLS 507 Introduction to Systems Engineering
ETLS 508 Systems Design
ETLS 509 Verification and Validation
ETLS 601 Program/Project/Team Management
ETLS 789 Modeling and Simulation

SEIS 610 Software Engineering
SEIS 635 Software Analysis and Design
SEIS 640 Operating Systems Design
SEIS 645 Computer and Network Communications
ETLS 880 Systems Engineering Project

Systems Engineer

ETLS 507 Introduction to Systems Engineering
ETLS 508 Systems Design
ETLS 509 Verification and Validation
ETLS 601 Program/Project/Team Management
ETLS 789 Modeling and Simulation

ETLS 671 Human Aspects of Technical Management
SEIS 601 Foundations of Software Development
ETLS 506 Statistical Methods for Manufacturing Quality
ETLS 672 Excellence in Design
ETLS 880 Systems Engineering Project

Engineering Project Manager

ETLS 507 Introduction to Systems Engineering
ETLS 508 Systems Design
ETLS 509 Verification and Validation
ETLS 601 Program/Project/Team Management
ETLS 789 Modeling and Simulation

ETLS 661 Eng Economic Analysis & Prod Control
ETLS 671 Human Aspects of Technical Management
ETLS 672 Excellence in Design
ETLS 701 Design of Experiments
ETLS 880 Systems Engineering Project

Medical Device Systems Engineer

ETLS 507 Introduction to Systems Engineering
ETLS 508 Systems Design
ETLS 509 Verification and Validation
ETLS 601 Program/Project/Team Management
ETLS 789 Modeling and Simulation

ETLS 520 Design & Manufacturing in Medical Device Industry
ETLS 722 FDA Medical Device Quality Systems
ETLS 724 Medical Device Clinical Studies
ETLS 731 Combination Products, Drugs and Biologics
ETLS 880 Systems Engineering Project

In addition to training and education, many organizations choose to use significant on-the-job training to develop their associates. This training may be as simple as providing stretch assignments to department members, but may also include things like structured assignments. With the structured assignment approach, a person is given a task that will help him or her to learn a specific knowledge area within the department. Structuring assignments from something straightforward and isolated to something more complex will help guide the associate through the development process, learning one piece at a time until larger and broader understandings are achieved. This method would suggest that a series of tasks be provided to the associate in increasing order of complexity:

Fix an error in an existing system
Modify an existing feature to extend it to multiple systems
Implement a new feature following a currently implemented design
Modify or implement a new feature using the new approach

With structured assignments the organization can begin to explore the level of understanding of the associate, and identify additional gaps that could be addressed as well as previously unknown skill areas that can be used for the benefit of the organization.

Another area to be explored by the Engineering leadership is the development and sustainment of a mentoring program. The function of a mentoring program is to pair an inexperienced individual from within the organization with an experienced individual in order to share knowledge and understanding about the operation of the organization. Often these mentoring programs tend to extend well beyond simply creating a familiarization with technical aspects of the program or project, and develop into career and life planning for associates involved. There are many benefits to mentoring, and although they vary from program to program and associate to associate, many factors are common to the majority of mentoring offerings.

For the mentor:

Allows the mentor to give back to the organization
Develops inability in the mentor to listen actively rather than passively
Encourages the mentor to share knowledge, which helps increase the mentor's self-esteem
Strengthens the mentor's interpersonal relationship skills
Teaches the mentor about other departments and functions within the organization
Helps refocus the energy of the mentor on the development of the organization

Increases job satisfaction for the mentor

For the Protégé

Increases the protégées' self-esteem
Protégé learns to take control of his or her career
Teaches the protégé how to speak up and be heard
Educates the protégé on how to accept feedback
Improves the protégées interpersonal relationship skills
Provides an important networking contact for the protégé
Helps the protégée better understand the organizational culture and political environment

For the organization:

Demonstrates to associates that the organization is willing to invest in them
Demonstrates externally that the organization values its associates
Fosters more loyalty among associates, reducing turnover
Creates a more inclusive work environment
Develops leadership skills in mentors
Encourages employee growth within the organization
Promotes a sense of cooperation and harmony throughout the organization

Chapter 15: Integrating the System

Any Systems Integration project can be either a nightmare or a blessing, depending on how well it is planned and executed. This becomes more the case if the project is extremely complex, or involves disparate applications and systems and has to be executed in conjunction with third parties. Under conditions such as these, it becomes extremely challenging to deliver the system within the desired timelines and at the required performance level. The integration of a complex system such as that described in this text is just such an application.

In order to accomplish the integration of disparate concepts, one might consider approaching this task as a System Development project. Thus, many of the concepts presented in the text can be directly applied in order to accomplish the integration of the system and processes within the Engineering organization.

We would propose the following methodology, describing a staged process to accomplish the integration:

1) Define a leadership and review team with members drawn from the organization and any significant external parties necessary for the implementation
2) Establish the project parameters, in terms of defining the purpose and corresponding and deliverables as well as any significant time constraints
3) Plan and implement a detailed project schedule, including interdependencies and resources required for completion
4) Define a team structure and engage the team within the project environment, and define what successful integration looks like in a common format
5) Establish the project priorities, in terms of selecting those methods and processes that the project will implements, as well as team communication needs and status for the project
6) Establish a business rhythm to ensure the project remains focused and on track for success

During the execution of the systems integration process, scope management will become a significant factor. Scope typically refers to all the work involved in creating the products of the project and the processes used to create them. A deliverable is a product produced as part of a project, such as hardware or software, planning documents, or meeting minutes. Scope management includes the processes involved in defining and controlling what is, or is not included in a project effort.

Scope management is the process by which we need to find a method of establishing the scope, defining the scope, verifying the scope, and controlling changes to the scope. Typically this might be accomplished with a project charter document, which will define the scope statement as well as deliverables defined for the project. Additionally, the work breakdown structure will contain many elements related to scope content.

As time progresses, the scope of the project should become clearer and more specific. It is likely to become more refined based upon the increased knowledge and insights gained in the development. This is shown in Figure 120.

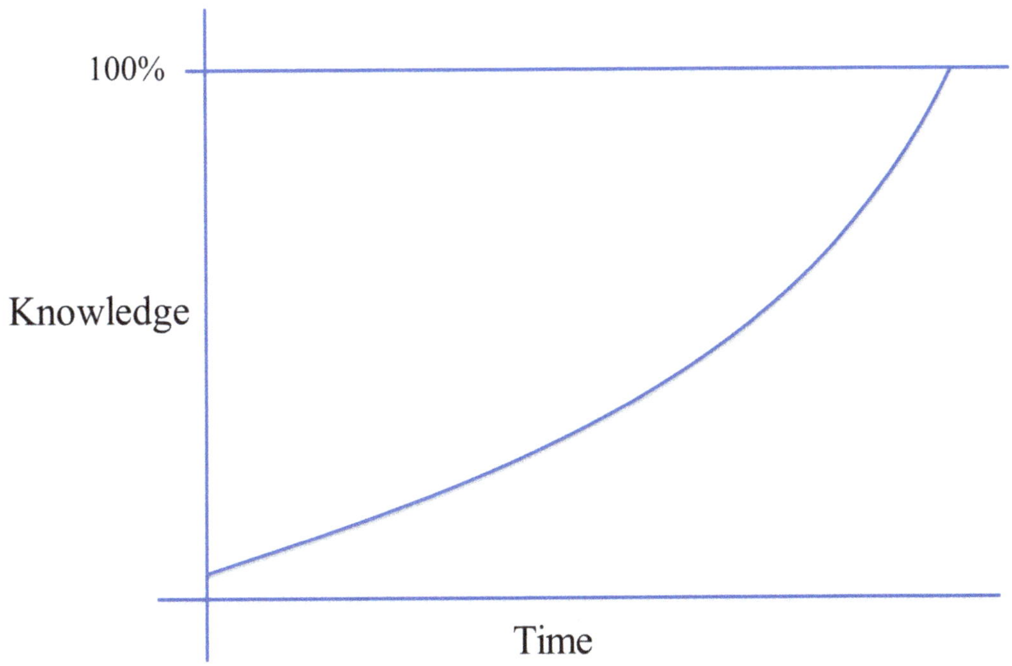

Figure 120: The Knowledge Curve

So why does this scope change significantly during the life of the project? First, it is important to recognize that the majority of planning for the project occurs very early in the lifecycle of the project itself. Second, the end product of the project is typically the generation of knowledge sufficient to allow the release of the project for production or process utilization.

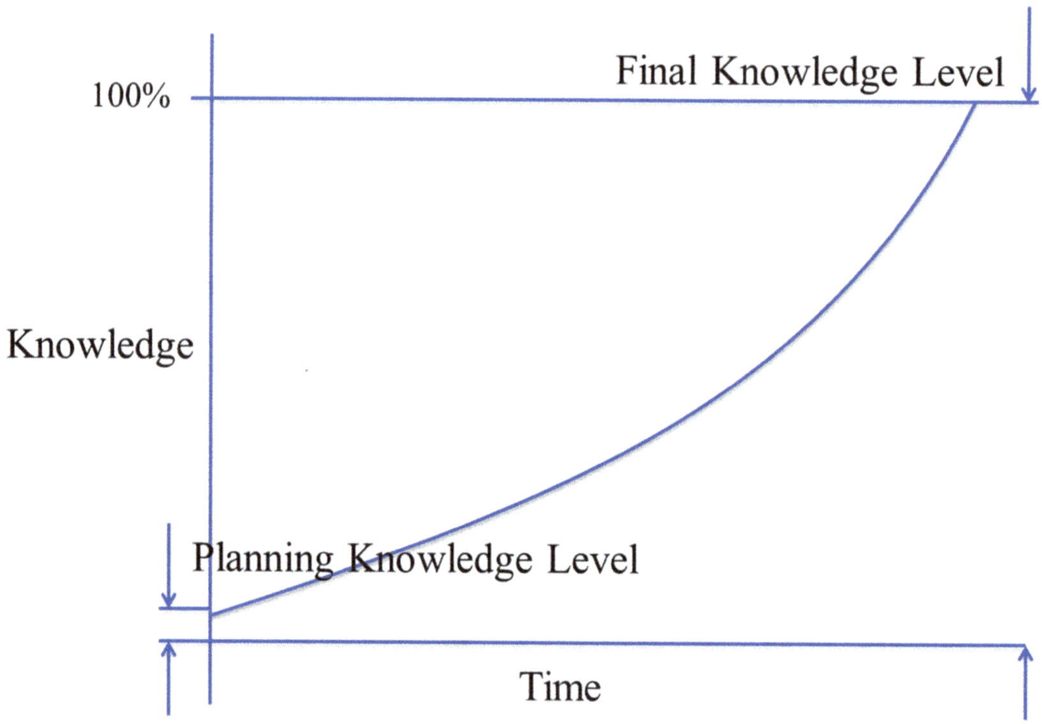

Figure 121: Why Project Scope Continually Changes

If one considers the level of knowledge developed at the outset of the project and compares it to the level of knowledge obtained by the completion of the project, it is apparent that many of the decisions and perspectives at the outset of the project may look less informed or totally incorrect by its completion. If this is true, it seems clear that in any system's development the Engineering leadership should fully expect significant scope changes. Therefore, scope control, which is often portrayed as the suppression of changes within the element, is actually the ability to appropriately manage scope change. This implies that the Engineering leader must be willing to accept the fact that projects and systems are exceedingly dynamic and will likely change on an ongoing basis during their development process. This is also true for systems integration efforts. As we become more familiar with the challenges and opportunities presented by the integration of the system of this complexity, we must be willing to constantly modify and adjust our plan in order to ensure that the deliverables are topical and appropriate for the work investment.

Of greater importance to the Engineering leader is effective change management. Change management is the process of developing a planned approach to implementing change in any organization. The objectives of change management are:

– Maximize collective efforts of stakeholders
– Minimize the risk of failure of implementation
– Ready the receiving organization for the results of our project

All projects incorporate change into your organization. Those that do not incorporate change are not necessary. By virtue of this, we must become active change agents within our organizations, as well as effective managers of the change as projects are implemented.

Although all people are unique individuals, we can have similar responses to change. In general, we view changes that are forced upon us as unnecessary and irrelevant. Many perceive the inconveniences caused by change as malicious attempts to sabotage our work or focus. In reality, the change implemented is rarely intended with ill will, and instead is simply not managed properly.

Organizational Change Management is the discipline of helping people adopt and accept change within the organization. It helps transition individuals, teams, and organizations from a current state to a desired future state. Why is change management important? Ultimately it enables timely project implementation in an engaged and ready workforce, and it avoids negative interruption to the business.

As an example, at GE Capital, one hundred percent of all projects evaluated as 'successful' had a good technical solution. Interestingly, over ninety-eight percent of all the projects evaluated as 'unsuccessful' also had a good technical solution.

Ultimately, change management involves the development of both stakeholder management, as well as change adoption processes within the organization. These two factors are defined as the quality and acceptance of the solution.

Stakeholder Management is the process by which a project's key stakeholders are identified and an approach to win or maintain their support is developed. Good Stakeholder Management leads to quality project execution and deployment.

Change adoption is intended to enhance the organization's readiness to accept change. The majority of this effort is to recognize that change in an organization requires time. In general, people will pass through varying reactions to the change as they process an implementation. These stages include:

Satisfaction
Denial
Resistance
Exploration
Hope
Commitment

Therefore, to manage the change adoption appropriately, we must follow a simple process. This includes:

Define why this change is needed
- Explain why the approach selected is optimal

Craft a business case
- Demonstrate or prove the business benefit

Create a compelling story for the change
- Make this story simple to understand and share

Align various stakeholders

- Focus on Consortium building
- Provide a constant flow of communications to get the organization accustomed to the change prior to the implementation

It is imperative for the Engineering leader to understand that all Engineering endeavors result in significant change introduced into the organization. The success of the endeavor is dramatically impacted by the individuals to whom the change was introduced. As an example, in developing a significant process improvement for the Manufacturing floor, you are introducing significant change to the operators who are interested in producing consistency and quality in their daily efforts. In reality, the individuals who have been negatively impacted by the introduction of a change are the same individuals who will either make or break the project. Therefore, it is imperative for the project leader to focus on the Change Management process well in advance of the implementation of the change.

References

Bak, P. (1996) How Nature Works. Copernicus (Springer). New York.

Christensen, Clayton M. (1997) The Innovator's Dilemma: When New Technologies Cause Great Firms to Fail. Boston, MA. Harvard Business School Press.

De Weck (2016) Multi-Project Program Management, Lecture 18 Slides, MIT Online Course. ESD.36

Gray, C. (2016) Program Management, a Primer. Online Whitepaper. www.softexpert.com

Peters, T. (1988) Thriving on Chaos. Harper Perrenial. New York.

Terninko, J, Zusman, A (1998) Systematic Innovation: An Introduction to TRIZ. CRC Press. New York.

Website (2015) IMP example. Planning and Scheduling Excellence Guide (PASEG). http://www.goaztech.com/paseg/

Website (2016) Coal Plant Schematic. https://en.wikipedia.org/wiki/Fossil-fuel_power_station

Website (2015) DODAF Product Guides. http://aitc.aitcnet/dodaf/

Website (2016) Systems Engineering Diagram and Guides. www.dau.mil

Website (2016) USDA Project Management Handbook, Version 1.0, online publication

Website (2016) Toaster Schematic Diagram. http://www.toaster.org/works.html

Website (2016) 25 Benefits of Mentoring. www.management-mentors.com/resources/benefits-of-mentoring

Wheelwright, et al (1984) Restoring our Competitive Edge. John Wiley and Sons. New York.

Vissers, M (2016) Merged Strategy image. http://marloesvissers.weebly.com/p2.html

Zimmerman, F (1997) Personal Interview. University of St. Thomas. Saint Paul, Minnesota

Appendix A – Verification and Validation Plan

SONIQ

ACOUSTIC SYSTEMS

VERIFICATION AND VALIDATION PLAN

Project or System Name

P.O.C.: Project leader

Month, Year

Revision Sheet

Release No.	Date	Revision Description
Rev. 0	08/15/15	V&V Plan release

Verification and Validation Plan

TABLE OF CONTENTS

NOTE TO AUTHOR: *Highlighted, italicized text throughout this template is provided solely as background information to assist you in creating this document. Please delete all such text, as well as the instructions in each section, prior to submitting this document.* ***ONLY YOUR PROJECT-SPECIFIC INFORMATION SHOULD APPEAR IN THE FINAL VERSION OF THIS DOCUMENT****.*

The V&V Plan provides guidance for management and technical efforts throughout the test period. It establishes a comprehensive plan to communicate the nature and extent of testing necessary for a thorough evaluation of the system. This plan is used to coordinate the orderly scheduling of events by providing equipment specifications and organizational requirements, the test methodology to be employed, a list of the test materials to be delivered, and a schedule for user (tester) orientation and participation. Finally, it provides a written record of the required inputs, execution instructions, and expected results of the system test.

1.0 GENERAL INFORMATION

1.1 Purpose

Describe the purpose of the V&V Plan. (Why are we doing the project and the plan?)

1.2 Scope

Describe the scope of the V&V Plan as it relates to the project. (What is included in he plan? What is excluded from the plan?)

1.3 System Overview

Provide a brief system overview description as a point of reference for the remainder of the document. In addition, include the following:

- System name or title
- Operational status
  - Operational
  - Under development
  - Undergoing a major modification
- System environment and special conditions
- Include a graphical representation of the system

1.4 Project References

Provide a list of the references that were used in preparation of this document. *Examples of references are:*

- *Previously developed documents relating to the project*
- *Documentation concerning related projects*
- *Standard procedures documents*

1.5 Acronyms and Abbreviations

Provide a list of the acronyms and abbreviations used in this document and the meaning of each.

1.6 Points of Contact

1.6.1 Information

Provide a list of the points of organizational contact (POCs) that may be needed by the document user for informational and troubleshooting purposes. Include type of contact, contact name, department, telephone number, and e-mail address (if applicable). Points of contact may include but are not limited to helpdesk POC, development/maintenance POC, and operations POC.

1.6.2 Coordination

Provide a list of organizations that require coordination between the project and its specific support function (e.g., installation, manufacturing, service, training, etc.). Include a schedule for coordination activities.

2.0 TEST EVALUATION

2.1 Requirements Traceability Matrix

Prepare a functions/test matrix that lists all application functions on one axis and cross-reference them to all tests included in the test plan.

Req	Description	Inspect	Analyze	Test	Demonstrate	Test #
1	System shall operate on 440AC	X			X	T123
2	System shall operate saw blade at 2500 RPM				X	T456
3	System shall have a safety factor of 2.0 on bridge		X			N/A
4						
5						

6						

2.2 Test Evaluation Criteria

Decide the specific criteria that each segment of the system/subsystem must meet. Such criteria are described by the user of the system/subsystem and typically are a mix of functional and performance requirements, such as processing data within a certain time frame, producing a report, or responding to an online query within a certain amount of time.

2.3 User System Acceptance Criteria

Describe the minimum function and performance criteria that must be met for the system to be accepted as "fit for use" by the user or sponsoring organization.

3.0 TESTING SCHEDULE

3.1 Overall Test Schedule

Prepare a testing schedule to reflect the unit, integration, system acceptance, and release tests, as well as the time duration of each. This schedule should reflect the personnel involved in the test effort and the site location. In the test schedule, include the following information:

- Documentation review
- Test scripts
- Data preparation
- Test execution
- Output review
- System certification
- System release
- Return of test site to pretest condition

3.2 Security

Prepare a list of requirements necessary to ensure the integrity of the testing procedures, data, and site. Any special security considerations (e.g., passwords, classifications, security or monitoring software, etc) should be described in detail.

3.3 Testing Location

This section provides a description of testing locations. Each location should be under a separate section header, 3.3 - 3.x Locations may include Park facilities as well as customer sites and facilties. Identify the location at which the testing will be conducted, and the organizations participating in the test. List the tests to be performed at this location.

3.3.1 Milestone Chart

Provide a chart to depict the activities and events listed below. When preparing the chart, give consideration to all tests scheduled for this location. The activities and events will be presented in chronological order with supporting narrative, as necessary, and will depict, for example:

- The overall on-site test period by calendar date, and portions of the period assigned to major portions of the test.
- The pretest on-site period required for system test team orientation, familiarization, and for system debugging.
- The period assigned for the collection of database values, input values, and other operational data required for system test.
- The period assigned for user training, operator training, maintenance and control group training, and management orientation briefing.
- The period assigned for preparation, review, and approval of the test analysis report.

3.3.2 Equipment Requirements

Provide a chart or listing of the period of usage and quantity required of each item of equipment employed throughout the test period in which the system is to be tested. Any communications and test data reduction equipment will be included.

3.3.3 Software Requirements

Identify any software required in support of the testing when it is not a part of the system being tested. Include systems support, communications, and applications software, their recording and storage media, version number, and media type.

3.3.4 Personnel Requirements

Provide a listing of the personnel necessary to perform the test. For each of the personnel, this listing should provide the following information:

- Name, title, current organization
- Description of the required tasks to be performed
- Geographical location of the work to be performed
- Time required (dates needed)
- Whether the requirement is full time, part time, or as needed
- Any special skills required (i.e., programming language, machine familiarity)

3.3.5 Deliverable Materials

Itemize all materials that will be delivered as part of the system test, to include the quantity and full identification.

3.3.6 Testing Tools

Identify the testing tools to be used during the preparation for and execution of the test.

3.3.7 Site Supplied Materials

Describe any materials required to perform the test that need to be supplied at the test site. These materials could include desks, chairs, special equipment, office supplies, database and its media, as well as other input and their media.

4.0 TESTING CHARACTERISTICS

4.1 Testing Conditions

Indicate whether the testing will use the normal input and database or whether some special test input is to be used.

4.2 Extent of Testing

Indicate the extent of the testing to be employed. Where limited testing is to be employed, the test requirements will be presented either as a percentage of some well-defined total quantity or as a number of samples of discrete operating conditions or values. Also, indicate the rationale for adopting limited testing.

4.3 Data Recording

Indicate data recording requirements for the testing process, including data not normally recorded during system operation.

4.4 Testing Constraints

Indicate the anticipated limitations imposed on the testing because of system or test conditions (timing, interfaces, equipment, personnel).

4.5 Test Progression

In progressive or cumulative tests, include an explanation concerning the manner in which progression is made from one test to another so that the cycle or activity for each test is completely performed.

4.6 Test Evaluation

4.6.1 Test Data Criteria

Describe the rules by which test results will be evaluated.

4.6.1.1 Tolerance

Discuss the range over which a data output value or a system performance parameter can vary and still be considered acceptable.

4.6.2 Test Data Reduction

Describe the technique to be used for manipulation of the raw test data into a form suitable for evaluation, if applicable. *The available techniques may include:*

- *Manual collection and collation of system test output into test sequence order, followed by verification of the results.*

- *Automatic inspection of test results as obtained by data recording means using a test data reduction program followed by manual inspection of selected test results which do not lend themselves to complete reduction by automatic means.*
- *Automatic inspection of test results specifically recorded for manipulation by the test data reduction program. Test results as recorded, include all items of test significance. The test data reduction program contains an image of correct data output for an item by item comparison of data, and provides a summary of an evaluated test as output.*

5.0 TEST DESCRIPTION

This section provides a description of the tests. Each test should be under a separate section header, 5.1 - 5.x.

5.x Test Identifier

Provide a test name and identifier here for reference in the remainder of the section. Describe the test to be performed.

5.x.1 System Functions

Provide a detailed list of the system functions to be tested.

5.x.2 Test/Function Relationships

Provide a list of the tests that constitute the overall test activity. Include a test/function matrix summarizing the overall allocation of the system tests to the functions.

5.x.3 Means of Control

Indicate whether the test is to be controlled by manual, semiautomatic, or automatic means.

5.x.4 Test Data

Identify any security considerations in each of the following subsections.

5.x.5 Test Procedures

5.x.5.1 Procedures

Describe the step-by-step procedures to perform each test.

5.x.5.2 Setup

Describe or refer to standard operating procedures that describe the activities associated with setup of the facilities to conduct the test, including all routine machine activities.

5.x.5.3 Initialization

Itemize, in test sequence order, the activities associated with establishing the testing conditions, starting with the equipment in the setup condition. Initialization may include functions such as:

- Queuing of data input values for the test
- Queuing of support software
- Coordination of personnel actions associated with the test

5.x.5.4 Preparation

Describe, in sequence, any special operations such as:

- Inspection of test conditions
- Instructions for data recording
- Modifications of the data base
- Interim evaluation of test results

5.x.5.5 Termination

Itemize, in test sequence order, the activities associated with termination of the test, such as:

- Recording readouts and critical data from indicators for reference purposes
- Termination of operation of support software and test apparatus
- Collection of system and operator records of test results

www.ingramcontent.com/pod-product-compliance
Lightning Source LLC
LaVergne TN
LVHW082118200926
848484LV00008B/1975